THE FUTURE OF CHANGE MANAGEMENT (VOLUME 1)

THE FUTURE OF CHANGE MANAGEMENT (VOLUME 1)

COLLECTED ESSAYS FROM LEADING THINKERS AND PRACTITIONERS

EDITORS:
PAUL GIBBONS AND TRICIA KENNEDY

Copyright © 2024 Paul Gibbons and Tricia Kennedy

All rights reserved.
(Please cite authors when using any material from this work.)

ISBN 979-8-9900855-0-3 (paperback version)
ISBN 979-8-9900855-1-0 (ebook version)

Publisher: Phronesis Media | Denver, Colorado, USA
Copy editor and proofreader: Tricia Kennedy
Diagrams: Andrés Goldstein and Tricia Kennedy
Cover design: Tricia Kennedy
Interior and typesetting: Priya Paulraj

All views, thoughts, and opinions expressed herein are the authors' own and do not necessarily reflect official policies or views of their employers or clients.

Product or corporation names may be trademarks or registered trademarks and are used only for identification and explanation without intent to infringe.

ALSO PUBLISHED BY PHRONESIS MEDIA

The Future of Change Management series

Book I: *The Future of Change Management: Collected Essays From Leading Thinkers and Practitioners* (volume 1, 2024)

Book II: *The Future of Change Management: Collected Essays From Leading Thinkers and Practitioners* (volume 2, 2024)

Book III: *The Future of Change Management: Collected Essays From Leading Thinkers and Practitioners* (volume 3, 2025)

Leading Change in the Digital Age series

Book I: *The Science of Organizational Change: How Leaders Set Strategy, Change Behaviors, and Create Agile Cultures* (2nd edition, 2019)

Book II: *Impact: 21st-Century Change Management, Behavioral Science, and the Future of Work* (1st edition, 2019)

Book III: *Change Myths: The Professional's Guide to Separating Sense from Nonsense* (1st edition, 2023)

Humanizing Business series

Book I: *The Spirituality of Work and Leadership: Finding Joy, Meaning, and Purpose in What You Do* (1st edition, 2020)

Book II: *Culture, Capitalism, and Sustainability: A Guide for Purposeful, Ethical Businesses* (due 2024)

PRAISE FOR *THE FUTURE OF CHANGE MANAGEMENT*

To implement the human side of strategic plans, CEOs need to consider what factors impede or enable change. The Future of Change Management unveils cutting-edge insights into human dynamics within organizations, making it essential reading for change experts, HR professionals, and executives.

DAVID BENNETT, CHAIRMAN, ALLFUNDS PLC

I've been following the "change must change" conversation initiated by Paul Gibbons some 10 years ago - this book will be a landmark. Too many change gurus are willing to play it safe - and advance tepid ideas, or rebrand old ideas as new. These conversations, about mental health, GenAI, and design thinking are JUST the ones we ought to be having.

WAYNE RESCHKE, FORMER CHRO ALLIANT ENERGY

Paul and Trica offer a delightful service by updating insights (theory, research, and practice) from many disciplines into the change field. From psychology, we learn new ways of thinking about cognition, assumptions, neuroscience, and other mindset issues. From economics, systems theory, and sociology emerge approaches to how people work collectively in organizations and communities. Each of these mindset and behavior sections of this volume will give those interested in upgrading change both breadth and depth for timely innovations.

PROFESSOR DAVE ULRICH, UNIVERSITY OF MICHIGAN

To become adaptive and competitive during uncertain times, leaders need to consider the micro, meso, and macro levels of change. This avant-garde book makes it clear that the future of change management means challenging business norms and adopting broader, more person-centered approaches.

**CARRICK BROWN,
SERVICE DESIGN AND DEVELOPMENT LEAD, AGILE CHANGE COACH**

We need to switch up our game, bring in some science, and embrace new ways of working as change practitioners. The Future of Change Management isn't just a conversation starter; it's a game-changer, challenging what we thought we knew. It dives deep into the science and tools that lead to genuine, impactful change.

DANIEL LOCK, PRINCIPAL, DANIEL LOCK CONSULTING

PRAISE FOR *CHANGE MYTHS* AND *THE SCIENCE OF ORGANIZATIONAL CHANGE*

Change Myths exposes how many of the methods change leaders rely upon as truths are often handed down pseudoscience; the book shows why change experts have a responsibility to revisit the evidence behind what they recommend.

MIKE ISKANDARYAN, WORKFORCE STRATEGIST AND AGILE CHANGE, MCKINSEY & CO.

Change Myths is brilliant. It challenged my thinking and beliefs, decades of organizational change practices, and how I interpreted my own experiences—a healthy reconsideration of the "truths" about change that will improve the field of organizational change.

WAYNE RESCHKE, FORMER SVP, CHRO ALLIANT ENERGY

Paul has the broadest, deepest, and most current knowledge and experience in change of any thought leader that I track, and I consider myself an avid learner. Paul has been "at the coal face" of change, doing the heavy lifting, and still managed to build and deliver real innovation that drives business value, which, for my money, is the real acid test of credibility. Paul has an uncanny ability to stretch across several adjacent domains to connect dots, spot incongruences, and showcase opportunities for moving forward. He breaks through old molds and offers new ways of thinking about and delivering change.

GAIL SEVERINI, PRINCIPAL, SYMPHINI CHANGE MANAGEMENT

Gibbons towers above business thinkers in the way that Drucker did in an earlier era. Even Drucker did not bring to business thinking the breadth of scholarship and originality of thought that Gibbons does.

ROBERT ENTENMANN, FORMER GLOBAL HEAD OF MARKETS, ABN AMRO

ABOUT THE EDITORS

Paul Gibbons

Paul Gibbons is the founding partner in Phronesis Media and a keynote speaker on mental health, GenAI and the future of work, and organizational change (of course.) Before that, he was a Partner at IBM Consulting — the Talent Consulting practice's thought leader and futurist on behavioral science, culture, leadership, and the future of work. He has previously advised PricewaterhouseCoopers (PwC), KPMG, and Deloitte on talent, culture, and leadership.

From 2010-2020, he was on the keynote circuit across five continents, speaking on the future of business: Humanizing business, leading change, culture change, AI ethics, and the future of work. During that decade, he was also an adjunct professor of business ethics and leadership at several U.S. business schools.

He previously authored five books, most prominently The Science of Organizational Change and Impact, the first two books in the Leading Change in the Digital Age series. Those books birthed the conversation about change mythology. The first volume of his Humanizing Business series, The Spirituality of Work and Leadership, was published in 2021.

After "experimental careers" in computer science, derivatives trading, economics, and neuroscience, Paul spent eight years as a consultant at PwC before founding Future Considerations. That firm became Europe's top leadership consulting firm working with Shell, BP, PwC, KPMG, Barclays, and HSBC on leadership, strategy, and culture change.

Paul is a fellow of the Royal Society of Arts, a hyperpolyglot, ranked a "top-20 culture guru," and CEO "super coach" by CEO Magazine. In 2000, he was elected to the U.S. Academy of Management Council, and he is a member of the American Philosophical Association, the American Association for the Advancement of Science, and the Institute for Business Ethics.

He lives in the Denver area with his two sons and enjoys competing internationally at mind sports such as poker, bridge, MOBA, and chess. (You can catch him on television from time-to-time.)

Tricia Kennedy (she/her)

Tricia is a both a partner in Phronesis Media and the founder and the principal consultant at Kennedy Consulting Services, LLC (TriciaK.com), which is a boutique organizational change and leadership consulting firm that helps businesses excel through inclusive, human-centered, and evidence-based strategies and practices. She is known for her inclusive and collaborative style, inquisitive and reflective approach, and humble commitment to challenging the status quo in a quest for continuous improvement.

With over 15 years of experience, she is a seasoned enabler and facilitator of organizational change who specializes in a holistic change approach and a creative blend of art and science from multiple disciplines to deliver sustainable results. She has advised and

ABOUT THE EDITORS

worked with various clients over the years, including Microsoft, Medtronic, BNSF Railway, Lexus, Sony Pictures Entertainment, and Goldman Sachs.

Tricia's journey to a career in organizational change and leadership started with stints in graphic and web design, copywriting, and public relations for the entertainment industry. She then followed her lifelong fascination with and passion for the human condition to graduate school, where she discovered organizational change and evidence-based practice.

She is based in the Dallas-Fort Worth metroplex in North Texas, where she enjoys spending time with her partner, cats, and family, laughing and finding humor in this crazy journey called life.

ABOUT THE AUTHORS

(alphabetical order)

Beirem Ben Barrah

Beirem is founder and CEO of Neurofied, a brain and behavior consultancy. As an entrepreneur, consultant, and speaker, he combines his interests in organizations, behavioral psychology, and technology. He frequently speaks on topics like change management, decision-making, learning & development, entrepreneurship, and neurotechnology.

He is Dutch-Tunisian, lives in Amsterdam, the Netherlands, and has many hobbies including martial arts, making art, devouring books, and exploring techs such as virtual reality.

Beierem is the co-author of chapter 5 with Philip Jordanov, an exploration of the benefits of collaboration between behavioral scientists and change managers for effective organizational change.

ABOUT THE AUTHORS

Newton Cheng

Newton Cheng is a husband and father, world champion powerlifter, a mental health speaker and activist, and is Director of Health + Performance at Google. His goal is to offer a different model of vulnerable leadership that inspires culture change around mental health in the workplace so that we can take better care of ourselves and each other.

He earned his BS in Electrical engineering and began his career working on Sony's Playstation 3. After this, he earned his MBA from UC Berkeley then joined Google, where he oversees a global portfolio of programs that support the health and wellbeing of Google's workforce.

Newton is the author of chapter 2, about mental health and its role in organizational change.

Dr. Ignacio Etchebarne

Ignacio is a culture and leadership transformation consultant dedicated to bridging the gap between academia and practical leadership. With 16 years of hands-on experience and a Ph.D. in psychology, he has authored numerous scientific articles and guides clients towards success by tailoring evidence-based strategies to their unique circumstances. As a co-founder of Hi—Human Insight, he is on a mission to integrate evidence-based insights into the everyday realm of leadership, teams, and organizational culture.

Dr. Etchebarne is the author of chapter 3, about neurodiversity and its role in organizational change.

Dr. Patrick Gallagher

Patrick helps organizations improve through people-centered change. He has helped multiple organizations, large and small, to make meaningful improvements to their culture and employee experience. He holds a Ph.D. in social psychology from Duke University and specializes in translating the latest and best research into practical tools for organizational change. A seasoned consultant, Patrick builds cross-functional teams to diagnose root causes and design effective interventions. He has published peer-reviewed research papers and professional presentations based on real organizational data, as well as blogs, articles and white papers.

Patrick is the author of chapter 10, about analytics and their role in organizational change.

James Healy

James is a principal and the global lead of Deloitte's Behaviour First offering, where insights from anthropology, behavioral economics, neuroscience, psychology, and sociology are practically applied to help organizations address their most critical challenges—including technology adoption, culture change, cybersecurity, and sustainability.

He has extensive experience leading behavioral, cultural, and technological transformations in global organizations, and has led projects in more than 60 countries on six continents (in industries including banking, insurance, mining, oil and gas, government, education, health, and aged care). He also hosts *The B-Word* (https://the-b-word.libsyn.com), a popular podcast featuring leading figures from the social and behavioral sciences to explore what it means to be human and how organizations can better understand and influence behavior.

James is the author of chapter 4, exploring the weakness of core values of organizational culture and how behavioral science can benefit organizational change.

Philip Jordanov

Philip is a cognitive neuropsychologist and head of training and consulting at Neurofied. He helps organizations such as Novo Nordisk, the Dutch Government, and Johnson & Johnson change and improve with insights from brain and behavioral science.

Philip can regularly be found playing gypsy jazz guitar with his brother on stage, or, with the same stage passion, guest lecturing at universities and international conferences.

Philip is the co-author of chapter 5 with Beirem Ben Barrah, an exploration of the benefits of collaboration between behavioral scientists and change managers for effective organizational change.

Robert Meza

Robert is the founder and director of Aim For Behavior (aimforbehavior.com), a consulting firm that works with clients on building products, services, and strategies with a human-centered and behavioral science approach. With a background in economics, innovation, and design, he thrives on sharing knowledge and uplifting those who wish to apply a behavioral lens to their work. He is known for co-creating solutions that work for customers and employees by combining innovation, design, behavioral science, and psychology.

He has spent the last 15 years advising and working with various clients over the years, including PepsiCo, Emirates, Visa, governments, and startups across continents.

Robert is the author of chapter 7, which provides a suite of tools from behavioral science for use in organizational change.

Hilary Scarlett

Hilary is an international speaker, consultant and author. Her work has spanned Europe, North America and Asia, and focuses on helping leaders in the private and public sectors to introduce change efficiently and effectively to their organizations. In particular, she designs masterclasses and workshops based on the belief that if we can understand our brains better, we can work with that knowledge to help improve both our wellbeing and performance at work. She works with neuroscientists in the UK and the U.S., bringing their work out of the lab and into the workplace in a very practical and accessible way. Hilary is a member of the British Neuroscience Association and the British Psychological Society.

Hilary is the author of chapter 1, about neuroscience and its role in organizational change.

Yves van Durme

Yves is the global Organizational Transformation leader for Deloitte Consulting. His recent book, *Change Can Be Child's Play*, is available in Dutch and in English. He is a high-performance squash coach which influences his focus on sustainable performance and high-performing teams.

Yves, along with Paul Gibbons, is the author of chapter 8, about the uses and abuses of design thinking in organizational change.

Natasha Young

Natasha is a managing consultant at IBM, passionately dedicated to helping clients with their business transformation. She studied psychology at both the bachelor's and master's levels (University of Bath and University of London, respectively). Known for leading engaged, collaborative teams to deliver at pace, she focuses on building strong relationships with her clients, understanding their needs, and empowering them to reach their full potential. She brings her passion for psychological theory and research to improve people's working lives and bring about real and valuable change to organizational performance.

Natasha is the author of chapter 9, which explores practical applications of ChatGPT (and other generative AI applications) to organizational change.

Scott Young

Scott is an independent educator, advocate, and advisor who is passionate about helping private sector organizations apply behavioral science ethically and effectively to their organizations. Most recently, he was head of private sector at the Behavioral Insights Team (BIT) North America.

He also has experience with BVA Nudge Consulting, and spent 20+ years leading Perception Research Services, a global shopper insights agency. The author of three books and over 40 published articles, his academic credentials include regular guest lectures at the London School of Economics (LSE), UPenn's Masters of Behavioral & Decision Sciences (MBDS) program, and the University of Chicago Booth's Executive Education program in behavioral economics.

Scott is the author of chapter 6, an exploration of behavioral science's role in human relations (HR) practice.

ACKNOWLEDGMENTS

Your editors would like to thank Dave Ulrich, who agreed at short notice to pen a foreword.

All the thanks are due the contributing authors and others who volunteered their time to submit chapter proposals, those who contributed volume 1 chapters, and those who are doing so for future volumes.

The books Science of Organizational Change and Change Myths and The Future of Change Management project have attracted superfans—we are incredibly grateful for each one of you. When our energy and enthusiasm flags, say mid-project, having you to pump us up from the sidelines is heartwarming. (We won't embarrass anyone by outing you here. You know who you are.)

Paul is, as always, most grateful to his parents (late 80s) and his teen sons (Conor and Luca, 19 and 14). Tricia is forever grateful to her amazing circle of family and friends.

Last, but foremost, our authors contributed tirelessly and passionately, some revising a half dozen times, some putting up with delays on the editorial side. If change management has a bright future, it is because of you and other people like you.

TABLE OF CONTENTS

Also published by Phronesis Media	*v*
Praise for The Future of change management	*vii*
Praise for CHANGE MYTHS AND THE SCIENCE OF ORGANIZATIONAL CHANGE	*ix*
About the editors	*xi*
About the authors	*xiv*
Acknowledgments	*xxi*
Table of contents	*xxiii*
Foreword	*xxvi*

Chapter 0
Introduction — 1

Chapter 1
Brains in Constant Change — 11

Chapter 2
Mental health in workplaces and in organizational change — 27

Chapter 3
Neurodiversity that Excludes? Tailoring Change Initiatives for Neurodiverse Stakeholders — 60

Chapter 4
Culture Change: Beyond Shared Values　　　　88

Chapter 5
Evidence-Based Behavioral Science in Organizational
Change　　　　113

Chapter 6
Applying a Behavioral Science Lens to
Human Resources　　　　129

Chapter 7
Behavioral Science Tools for the Change Professional　　144

Chapter 8
Uses and Abuses of
Design Thinking　　　　168

Chapter 9
Will ChatGPT Replace the Change Manager?　　　　182

Chapter 10
People Analytics Accelerates Change　　　　203

Chapter 11
Conclusion　　　　226

Appendix I
Contact the contributing authors　　　　230

References and Further Reading　　　　**232**

FIGURE LIST

Figure I.1: Sampling of factors driving the future of change management.
Have today's change models kept up with the latest science and workplace trends? We think not.   3

Figure I.2: The Future of Change Management Volume 1
Change management will look different in 2030. Do we know how? What will be included?   8

Figure I.1: Threat-reward states and performance.
We can lead more effectively when we recognize threat-reward states in ourselves and team members.   15

Figure I.2: Inverted U of performance.
There is an optimal level or arousal—leaders can ask themselves where they and their followers may be on this curve.   18

Figure I.3: The ACCESS framework is a memorable summary of brain care for people.
Leaders can use the ACCESS framework on themselves or with their teams.   23

Figure II.1: Famous sufferers from depression.
Depression not only affects some celebrities, but also affects more than a billion people worldwide.   29

Figure II.2: Externally, Newton was winning medals; internally, he was suffering.
Mental health affects high performers. Visible success can mask inner distress.   31

Figure II.3: Patient health questionnaire (PHQ-9).
PHQ-9 is a way to recognize the symptoms of depression in yourself and perhaps others. It is less threatening than longer, older depression questionnaires.   54

Figure II.4: Guidance for change professionals, managers, and leaders.
Change managers can alter some of their communications and training and, critically, coach and support managers to be mental health aware during their role.   57

Figure III.1: Badging, disability, or different abilities?
Do labels and badges help? Focus on deficit labels conceal strengths.   67

Figure III.2: Neuropsychiatric perspective of health and illness.
The psychiatric, or diagnostic paradigm of the medical profession classifies people bluntly, which overlooks performance strengths.   68

Figure III.3: Neurodiversity movement perspective.
Neurodiversity contains variation where difference does not always equate to a deficit. Who gets to decide which is which?   69

Figure III.4: Neurodivergent strengths.
People with neurodiverse labels have underrecognized strengths.   71

Figure III.5: Specialist vs. generalist cognitive profiles.
Again, neurodivergent individuals comprise deficits and strengths despite a traditional focus on deficits alone.   72

Figure III.6: Spectrum of neurodiversity.
Neurodiversity can be visible or invisible and invisible neurodiversity is equally important as visible.   74

Figure III.7: Executive functions of the brain.
Imaging research localizes executive-level functions in the prefrontal cortex in humans, but this finding can mask the diversity of capability.   76

Figure III.8: Neurodiversity Spectrum Assessment (Hi-NSA®) for STEERing leaders and teams.
Build self-awareness of your own executive-function capabilities using HI-NSA®.   79

Figure III.9: Four brain-based change strategies.
Development strategies should take multiple factors into account, including cognitive diversity.   81

Figure IV.1: Sampling of corporate values from Fortune 500 companies.
Figure caption. Businesses talk about values non-stop and very different businesses claim similar values. But do they mean anything?   92

Figure IV.2: McKinsey 7S model.
More than 40 years ago, McKinsey began to tout the importance of the softer side of organization. The 7S model has stayed with us since.   94

Figure IV.3: The Inglehart-Wetzel World Cultural Map.
World cultural values appear to be diverging, not converging.   99

Figure IV.4: A plastic fly seems to solve the millennia-old problem of improving men's aim.
A fake fly in the urinals at Amsterdam's Schiphol Airport was enough to reduce overall cleaning costs by 8%.   106

Figure IV.5: Salient switches—timely, salient nudges work better.
Timely nudging for behavior change is more effective than educating or berating.   107

FIGURE LIST

Figure IV.6: Walk this way.
Figure caption. The COVID-19 pandemic introduced a whole range of nudges, many of them applying Richard Thaler's aphorism, "If you want people to do something, make it easy." **108**

Figure V.1: A taxonomy of evidence-based change interventions in six categories.
Evidence-based change interventions are useful across all aspects of change management. **121**

Figure VI.1: Critical questions to answer during behavioral change analyses.
Many HR initiatives fail because the essential behaviors are not clearly specified. **132**

Figure VI.2: EAST stands for easy, attractive, social, and timely.
Using EAST, behavioral interventions can be designed efficiently for maximum effect. **133**

Figure VI.3: TESTS stands for target, explore, scale, trial, and solution.
TESTS is an efficient checklist for building measurement into behavioral science interventions. **141**

Figure VII.1: Taxonomy of behavioral-change techniques (BCTs).
Michie's 2013 taxonomy of 93 behavioral change techniques. Yes, there are so many it is barely legible, this chapter is here to help. **147**

Figure VII.2: Behavior Change Wheel (BCW) from Michie's team at University College London (UCL).
BCW helps you work inside out starting with the challenge, then understanding which behavior(s) to focus on, what the barriers and enablers are (drivers) stopping or enabling a behavior to occur, then from there you could select broad intervention types (and policies) and then getting to the nitty gritty of how BCTs can be used to address the drivers. **149**

Figure VII.3: A simplified version of a behavioral systems map.
Using a system's map means effort isn't wasted and is directed toward the highest leverage points. **152**

Figure VII.4: Define behaviors as specifically as possible.
Defining behaviors very specifically is an essential early step designing behavioral interventions. **153**

Figure VII.5: COM-B's components and their definitions.
COM-B analyses the drivers of behavior by taking data and looking for patterns or clusters. **155**

Figure VII.6: COM-B in workshop format for use with stakeholders.
COM-B is simple and intuitive enough that it can be used in workshops, to gain stakeholder perspectives on the drivers of behavior. **156**

Figure VII.7: COM-B and the Theoretical Domains Framework
The TDF provides a more granular look at the drivers of behavior. **158**

Figure VII.8: Theory and techniques (T&T) tool in all its complexity.
Once you decide what approach you will take to changing the specific behavior, you want to select an approach supported by evidence. **161**

Figure VII.9: COM-B and theoretical domains framework (TDF).
We cross-reference an intervention with a TDF area such as skills. **162**

Figure VII.10: Final output of the T&T tool.
Using the T&T tool allowed us to design and prototype two kinds of interventions. **164**

Figure VII.11: APEASE tests for practicality.
Simple checklists like APEASE are useful for making solutions fit. **165**

Figure VII.12: APEASE allowed Talent Builders to examine many implementation issues often ignored.
We used APEASE in workshop format at Talent Builders to great effect. **166**

Figure VIII.1: Influencing stakeholders after the fact can require coercion, sometimes extreme.
In Game of Thrones, Queen Danaerys used Drogon to influence key stakeholders in her preferred direction. **170**

Figure VIII.2: Design thinking process.
Design thinking is an iterative process that builds in creativity, empathy, engagement, and learning from the get-go. **172**

Figure VIII.3: Where can we use the design mindset in business?
From its cool-kids roots, design thinking has become the go-to tool for high engagement design and problem-solving processes. **175**

Figure VIII.4: Parisian stakeholders storm the Bastille.
Involve stakeholders or risk lack of alignment and even antipathy toward project goals. They might even storm your castle (Prise de la Bastille, Anon, 1790). **178**

Figure VIII.5: Figure title. Design thinking, Lean startup, and Agile diagram.
Full stack designers can integrate design thinking, Lean, and Agile, but must use change management to drive the entire process. **180**

Figure IX.1: ChatGPT took only months to achieve what other 'exponential' apps did in years.
Even faster than TikTok? It took Uber six years to achieve what took ChatGPT only two months (data from Statista). **183**

Figure IX.2: GenAI use cases are proliferating, a few examples from just six months after launch.
But can it cook breakfast? As the world tinkers with GenAI, use cases continue to expand. **186**

FIGURE LIST

Figure IX.3: Organisational change use cases for generative AI (e.g., ChatGPT).
It is early days for ChatGPT and organizational change, so far, it has proven useful.   **188**

Figure IX.4: ChatGPT doesn't evaluate validity, only popularity.
If you used ChatGPT in 1400, it would tell you that the world is flat; it is limited by the veracity of the internet.   **197**

Figure IX.5: ChatGPT has naïve view of change resistance.
With its concise answers, ChatGPT risks oversimplifying.   **198**

Figure IX.6: Useful list of talking points returned when ChatGPT is asked to role play a conversation about change resistance.
The change manager can role play tricky conversations with stakeholders.   **199**

Figure X.1: Five dimensions of change metrics
Change metrics help change leaders know whether what they are doing is working, and how they might have to adjust.   **206**

Figure X.2: Illustrative output of organizational network analysis (ONA).
An analysis of the frequency and volume of email, Slack, and other communication channels reveal patterns like those depicted. Darker circles represent people who are particularly central or could potentially bridge silos, making them strong candidates to help lead change adoption.   **216**

Figure X.3: Six tips for using people analytics to aid change efforts.
Enlisting people analytics to help with change efforts may not be easy, but the change leader who follows these steps may gain efficiency and effectiveness.   **220**

Figure XI.1: The Future of Change Management (volume 2).
Candidate topics under consideration for the next, or second, volume in The Future of Change Management series.   **229**

FOREWORD

DR. DAVE ULRICH
Rensis Likert Professor,
Ross School of Business, University of Michigan

Decades ago when I did my Ph.D. second-year exam, I was asked to write an essay on the future of organization development (OD). I do not recall what I wrote, but I do remember the title, "When, why, and how will OD OD?" I remember even more with angst and amusement (now) that I was told this setting was not the time for creativity, but rigor and scholarship, and that my committee would determine my seriousness in six months (i.e., I failed).

Since then, I have followed, explored, and practiced the work on "change" that has gone through many name evolutions: Adaptation, agility, alignment, blur, change, fast, fitness, invention, OD (it has not OD'd), rapid, reinvention, reengineering, renewal, switch, tipping point, transition, transformation, urgency, velocity, and others. As I review the intent and content of this excellent volume, I ask myself three questions similar to those in that decades-ago exam.

Why (what's so)? No one can question the increased pace of social, technological, economic, political, environmental, and demographic changes that touch countries where we live, companies where we work, and personal lives where we socialize, play, and worship. What's so? Change is happening all around us. Why does this matter? Because countries prosper, organizations thrive, and individuals flourish based on their ability to manage change.

What (so what)? As Paul and Tricia wisely note, the "content" of change changes as new ideas and approaches occur about what to change. I should express a bias about the content of change that is often framed as a from-to logic. Our (and other research) shows that successful organizations, leaders, and individuals *navigate* rather than manage paradoxes. This means that instead of moving from A to B to C to D, success means continuing to do A and also B and also C and also D. Navigating paradox builds on the past to live today and create tomorrow. Within the last year, I re-read *Management and the Worker* (1939 version), and classics by Douglas McGregor, Kurt Lewin, March and Simon, and Peter Drucker. These amazing thought leaders built a solid foundation for change that so many others should acknowledge and build on.

As approaches to change build on their timeless ideas, timely innovations will emerge given the new context of business. None of these authors would write today what they wrote then. Paul and Trica offer a delightful service by updating insights (theory, research, and practice) from many disciplines into the change field. From psychology, we learn new ways of thinking about cognition, assumptions, neuroscience, and other mindset issues. From economics, systems theory, and sociology emerge approaches to how people work collectively in organizations and communities. Each of these mindset and behavior sections of this volume will give those interested in upgrading change both breadth and depth for timely innovations.

How (now what)? For most change agents as researchers, consultants, practitioners, or leaders, turning what we know into what we do is critical. Ideas without impact are ivory tower daydreams; impacts not grounded in theory are irreplicable sometimes random events. Many of the tools for making change happen that have been around for some time continue to have efficacy. I still use Lewin's force field analysis to evoke conversation and where we are, where we want to go, and how to manage drivers and restrainers to make progress. The Hawthorne studies offer foundational insights about the importance of engaging others in the change process. McGregor's Theory X and Y assumptions show up in multiple books on how leaders should change.

However, change tools advance to fit the times. Design thinking, for example, builds on systems theory to offer fresh ideas on making innovation happen. Analytics and access to data, not only through observations and surveys, but through GenAI applications enables information about change to build on benchmarking, best practice, and predictive analytics to offer guidance about what an individual, leader, and/or organization can accomplish changes that matter to them.

Not the conclusion. I like to end most of my workshops or talks by asking participants to think about "name the best year of your life." I often hear about youthful delight, relationships (e.g., marriage, kids, and sometimes divorce), accomplishments, and so forth. Then I remind the group that I believe the best year of your life should always be the next 12 months. We can look back to learning from good and bad experiences, but we can envision forward to shape opportunities. Likewise, GenAI (e.g., ChatGPT) helps us summarize the past quickly and brilliantly, but our creativity will discover a yet unknown future.

The same logic applies to the why, what, and how of change. We honor the past (even our own) and cherish facing an unknown and to-be-defined future.

This future focus on change by Paul, Tricia, and thought-leading co-authors will fold that future into our present. I wish had written this in my exam answer so long ago, but I believe I have and will also continue to learn!

CHAPTER 0

Introduction

by Paul Gibbons and Tricia Kennedy

"The old must always make way for the new, and one thing must be built out of the ruins of another."

LUCRETIUS (CIRCA 55 BCE)

Journey with us on a thought experiment. Imagine 10 years hence, a few dozen change experts gather to explore, "What has changed in change management over the preceding decade?" What would they say?

This might lead to other questions: Which change ideas have been discarded? Which new methods have been incorporated? Have old myths fallen into disuse, or do we still hear about burning platforms? How might artificial intelligence (AI) have been deployed? What, from the frontiers of behavioral science and organizational psychology, would have changed our understanding of how people and organizations change? Is change still conceived as a step-by-step, paint-by-numbers process? How has the future of work affected how organizations change?

And so on.

Perhaps not even fit for today?

Imagine the same change experts sitting down today to explore, "What do you think is the most important factor in organizational change?"

They might object to the question saying change is too complex to be reduced to one magic bullet, but if pressed, might offer some of the following: Trust, psychological safety, empathy, creativity, changing behaviors, leadership coaching, "lean change," emotional intelligence, agile teams, changing culture, facilitating conflict, and leadership alignment. And many more.

This list does not begin to capture the wealth of ideas from today's leading practitioners.

Now pull down your dusty old CMBOK, Kotter, Prosci, Conner, CCMP, or other change "bible." Find the pages on how to create trust, empathy, or creativity in them. Find psychological safety or facilitating conflict.

The old models ain't got it. We think not only that they aren't fit for 2030, but they also omit much of what change experts might name as the most important ideas.

When most change models were cooked up, in the 1990s, cell phones looked like bricks, email was a new, exciting technology, we browsed the internet using Netscape, and managed our time with Filofax.

Certifications have their place, but there is an inbuilt conservatism. Our review of three of the most popular ones was like jumping into a DeLorean with the Beastie Boys playing on the cassette player.

INTRODUCTION

Those of you familiar with our previous book, *Change Myths* (2023), may remember our review of how **the context** in which change management happens has changed.

There are three domains:

- The cultural and demographic context,
- The human sciences, and
- The business/ technological context.

Figure 0.1 illustrates just some of the factors in each domain.

FACTORS DRIVING THE FUTURE OF CHANGE MANAGEMENT

Cultural and demographic context	Human sciences	Market, business, and technology trends
Remote working	Beahvioral science	Analytics
Talent shortages and employee power	Habit science	Data-driven decision making
Global workforces	Well-being and mental health research	Digital transformation and digital adoption
Diversity and inclusion	De-biasing	Cloud
ACTs (Slack, Jira, etc.)	Evidence-based management	Generative AI and LLMs
App, message overload	Growth mindset	Globalization
Values polarization	Climate and psychological safety	Blockchain, metaverse, and crypto
#metoo, anti-racism, and backlash	Mindfulness research	Contingent workforce
Work and family balance	Purpose and meaning	War for talent
Learning preferences (bite-sized, video)	Cognitive-affective neuroscience	Enterprise agility
In-person training death	Choice architecture and nudges	Supply chain disruption and reshoring
Democritization and SMWT	Positive psychology and flourishing	Sustainability

© Phronesis Media • Future of Change Management, Gibbons & Kennedy

Figure I.1: Sampling of factors driving the future of change management.
Have today's change models kept up with the latest science and workplace trends? We think not.

Those early organizational change pioneers could not have imagined today's scope of remote working, inclusion, social media, psychological safety, growth mindset, AI, sustainability, and/or analytics.

Your editors applaud the pioneering spirit that codified the best knowledge of the 1990s into usable tools for change leaders. We also feel that the profession is evolving far too slowly—that is, change needs to change.

The Future of Change Management series

None of us know how the world will have changed a decade from now, but we feel certain that to continue to be relevant, change management must catch up and then evolve with the shifting business context and take advantage of discoveries in the human sciences.

What could your editors do to outline the newer ideas, the frontier, the contours of the field that would nudge the profession forward?

We, of course, have our own ideas on how this future might look, but we decided to harvest the knowledge of the change community around the world, for we are confident that the formal knowledge (in the books, certifications, and frameworks) hasn't kept up with the tacit knowledge (in the heads of practitioners.)

We got busy and did some research.

Our research, interviews, and surveys reached practitioners and academics on all the standard for a like LinkedIn, Medium, and Substack. We asked our IRL communities—colleagues and clients.

What came back were sixty or so topics.

There was some chaff with the wheat. Some were not so new. Some people said "Agile," which was birthed in 2001. An "expert" from a name-brand change consulting firm offered "capability building" and "increasing change management awareness." Not new.

Some offered ideas that were not robust in our estimation— and we have a bias toward topics for which there is evidence beyond the anecdotal.

Still, we were inundated with suggestions and chapter offers!

As soon as we sat down to write the first volume, it was clear that at least a second would be required, and likely a third. But that gets us ahead of ourselves.

Even with that input, in a vast field full of diverse views, we cannot advertise this collection as a map of the whole territory. But we need to start somewhere.

What is in volume 1?

On some topics that we wanted for volume 1, such as **diversity and inclusion**, and **sustainability**, we didn't find an author in time. We also have nothing on the burgeoning commercial area of change management software. There aren't chapters on systems thinking, habit change, Agile, change leadership, psychological safety, emergent change, complexity, or integral change. Yet. (Watch this space.)

We had to pick who went first and chose who we thought was best qualified. We also went for geographical diversity because (unsurprisingly) different continents approach change differently.

We didn't (this time) get anyone from Asia or Antarctica but have contributions from the other four continents. We even considered a multi-language book, with German, French, and Spanish chapters—but chickened out with the safer English option.

We organize the book into three sections: Minds, behaviors, and tools.

The chapters can be read in any order; each is "a complete thought" as it were.

In the minds section, our authors peek under the hood at what we are learning about change at the micro level—what we know about how the mind and brain function and that knowledge's effect on organizational change theory and practice. **Hilary Scarlett**, who wrote the change bestseller *Neuroscience for Organisational Change*, summarizes some of the most interesting advances in neuroscience and change. Then, **Newton Cheng**, Google's director of health and performance, has a dialog with Paul that illuminates some of the biggest issues in mental health and change, and highlights Google's efforts in that area. Then, to round out that section, **Dr. Ignacio Etchebarne** offers his research-based view on neurodiversity and change.

Those of you familiar with our earlier writing will know that we think developments in behavioral science have the potential to progress our field more quickly than perhaps any other. But we still get much confusion. One change expert with well over twenty years' experience from England said—"behavioral science is just psychology, innit?" Another, from the U.S., said, "Don't you mean behavioral economics?" Though the term behavioral science is hotly debated among behavioral scientists, this is our working definition:

INTRODUCTION

Behavioral science is an interdisciplinary field comprised of the behavioral components of the following sciences: Economics, decision science, psychology, neuroscience, systems thinking, sociology, and anthropology.

To begin the behaviors section, Deloitte Australia Principal **James Healy** offers his take on behavioral science and culture change—and how the values-led approach offered by specialist culture change consulting firms is flawed. **Philip Jordanov** and **Beirem Ben Barrah** from Neurofied then illustrate how change managers and behavioral scientists can collaborate to enable effective, lasting organizational change. Lastly, **Scott Young**, former principal and head of private sector at the Behavioral Insights Team, offers a case-study-rich look at how behavioral science complements change management and HR policy implementation.

Finally, in the tools section, **Robert Meza**, principal at Aim for Behavior, shares with us several behavioral science tools that he has developed to enrich the practitioner toolkit. Then **Yves van Durme**, Deloitte's organization transformation offering leader, shares a chapter with case studies on Deloitte's use of design thinking in change. The astute reader may scoff, "design thinking is from the 1970s, how is that new?" It isn't, but its use as a change methodology is new—and major consulting firms, such as IBM and Deloitte, have invested substantially in augmenting and deploying it. Following that, **Natasha Young**, managing consultant at IBM, shares her early experiments using ChatGPT as a work-a-day change management lead consultant. Finally, **Patrick Gallagher**, senior director of employee experience at Partners Health Management, shows how people analytics can support change.

CHAPTERS IN FUTURE OF CHANGE MANAGEMENT - VOL. I

MINDS
- Brains in Constant Change
- Mental Health in Workplaces and in Organizational Change
- Neurodiversity that Excludes? Tailoring Change Initiatives for Neurodiverse Stakeholders

BEHAVIORS
- Culture Change: Beyond Shared Values
- Evidence-based Behavioral Science in Organizational Change
- Applying a behavioral science lens to human resources

TOOLS
- Behavioral Science Tools for the Change Professional
- Uses and Abuses of Design Thinking
- Will ChatGPT Replace the Change Management Consultant?
- People Analytics Accelerates Change

© Phronesis Media • Future of Change Management, Gibbons & Kennedy

Figure I.2: The Future of Change Management Volume 1
Change management will look different in 2030. Do we know how? What will be included?

That is just the beginning, your editors have begun further discussions with thought leaders around the world about contributing to a second volume. In this book's conclusion, you will find a table of possible chapters for that second volume, likely due out in late 2024 or early 2025.

Editorial treatments

Each author has considerable expertise in their subject area. Your editors made suggestions, but even if we felt strongly otherwise, the expert authors had the final say in their chapter content.

INTRODUCTION

As one of our Irish grandmothers used to say, "You don't keep a dog and bark yourself."

We've also not insisted on uniformity of language. Some authors write in British English, others American. We left that alone. Some authors cited dozens of references. We've pared these for readability—but if you have a question or would like to discuss possible projects with our authors, their contact details are found in Appendix I.

MINDS

CHAPTER 1

Brains in Constant Change

by Dr. Hilary Scarlett

Though neuroscience burst onto the organisational change scene about 15 years ago, it would be a mistake to say that change management practice is "brain informed." Your editor believes that a) understanding and application of neuroscience is still patchy, and b) those who do apply it are often guilty of "brain overclaim"—making statements that aren't yet justified by science or are just plain wrong. One so-called change management expert told Paul, in all seriousness, that fMRI scanners could "read your thoughts."

Hilary makes none of these mistakes. Her book, Neuroscience for Organisational Change, is—for our money—the most robust one on neuroscience and change currently available. In this chapter, she maps out the territory and use cases of neuroscience's application to change management practice.

Adaptable or resistant?

Not long ago, organisational change had a beginning, a middle, and end. We talked about change 'programmes.' Those days have gone, as many of us have experienced. Change is constant. There are waves and waves and layers upon layers of change. It has become a way of 'being' at work. The challenge now is not about bracing for the next change programme and then heaving a sigh of relief when it seems to be over, it's about enabling employees to

develop capabilities and the mindset for constant flexibility.

Technology will take care of many of the more mundane tasks at work, what organisations need from employees is not only adaptability but also creativity, collaboration, and an ability to work with customers in a personal way.

The good news is that we humans can be extremely adaptable and resourceful. Just looking back to early 2020, when the Covid 19 pandemic arrived, organisations and employees moved incredibly quickly to adapt. Those who could work from home were equipped to do so within hours. The great majority of people changed the way in which they lived their day-to-day lives. In the bigger picture, people have adapted to living in every part of the planet, creating the clothing, buildings, heating, tools, lighting, etc. that enable us to live in a wide variety of climates, and now there are even people living in space.

So how can human beings be so adaptable in certain situations and so unwilling to change in others? As one member of a technology team in an investment bank asked, 'How come at home people want the latest phone or smart speaker, but when it comes to changing IT systems at work, they don't like it?'

Understand your brain

For most people, the brain is our key work tool. Just as we teach maths in school and how to use technology at work, we should teach everyone how their brain works, not least leaders. We don't need to become neuroscientists, just some basic understanding of what causes anxiety or stress at work, and also what brings joy, adaptability, and the ability for our brains to work at their best, would be useful.

Useful basics

Before taking a look at what helps and what gets in the way, it's useful to set out a few key points about our brains that shed light on why organisational change can be difficult.

Our brains are not designed for the 21st-century workplace

Our brains have not changed that much since our ancestors were on the Savannah. Whether we are at work or on the Savannah, the key goal for the brain is survival: Your brain just wants you to make it safely through the day. To achieve this, the brain wants to do two things: Avoid threats and seek out rewards. By far the more important of these two is avoiding threats: We can do without rewards such as shelter, food and even water for about a week and still survive: But if the sabre-tooth tiger gets you, you're dead. So, our brains are much more interested in threats than they are in the positives. Imagine for a moment that you are walking alone down a dark alley at night and you hear footsteps behind you. For most of us that would make us feel anxious, and yet most of us have never been attacked in a dark alley. From a survival point of view, it's better that we are anxious and alert than that we are too relaxed and vulnerable. Our brains lead us not to see the world as it is but in a way that is more likely to keep us safe. Anxiety and stress are the price we pay for the brain's safety-first agenda. For example, it is not just difficult meetings that stress us out the anticipation of such meetings also induces stress.

Threat and reward states affect our ability to perform

Being in such a threat state is not only unpleasant, it also affects our ability to perform. The 'fight or flight' response in the brain sends resources to those parts of the brain that help us to fight or run away, diverting them from the prefrontal cortex (just behind your forehead), that is often referred to as the executive centre of the brain because it is so important in decision making, emotional control, and working memory—abilities that are essential for work. When we are in a threat state, we cannot think at our best.

Threat states alter how we see the world, limit our thinking, reduce our energy, narrow our focus, and lead us to fall back on old habits (see Figure I.1). We are not so good at problem-solving, are less creative, less able to learn, and tend to see the world of work as more hostile than it really is: It affects our ability to remember and, over time, weakens our immune system. Not good.

When, on the other hand, we are in a reward state, the brain's 'seeking system' is activated. We are more curious and open to new ideas: We can be more creative, we can use our energy to explore and to connect with others because we don't need it for self-protection. The reward state, as its name suggests, feels good.

THREAT-REWARD STATES AND PERFORMANCE

Threat	Reward
Defensive, movement away	Thrive, movement toward
Mental energy protects	**Mental energy connects and creates**
• 'Fight or flight'	• Positive
• Think less clearly	• Focused
• Distracted and anxious	• Resilient
• Hypervigilant	• Collaborative
• Reduced memory	• Open to learning and new ideas
• Regress to old habits	• Creative and innovative
• Narrow minded	• 'Seeking system' activated
• Stress and high cortisol	• More able to empathize

Source: Scarlett, H. (2019). Neuroscience for organizational change: and evidence-based practical

Figure I.1: Threat-reward states and performance.
We can lead more effectively when we recognize threat-reward states in ourselves and team members.

Neuroscience can be applied at two levels, the organisational and the individual. We need to understand:

- What organisations need to do to create an environment that supports fast adaptation
- What individuals can do to help build our own ability to deal with constant change

Ten Things Organisations Can Do

There are many actions organisations can take to create a workplace that will nurture flexibility, let's take a look at a few.

1. Reduce the threat state at work

Most of us much of the time are probably in a threat state at work. Before the working day even starts, we have challenges about getting that child happily off to school on time, we wonder whether congestion on the roads or delays on public transport might make us late for work. If we are working remotely, technology might not be doing quite what we anticipated. Then once the working day starts, we often feel overwhelmed by the sheer volume of work, shifting priorities, having to work with people we don't particularly get on with, uncertainty about what's coming up, and not feeling our work is getting the appreciation it deserves. All this is against a backdrop of concern about climate change, the cost of living, conflict in various parts of the world, and so on. In a world where change is accelerating and constant, this could lead to people being in an even stronger threat state more of the time, hampering their creativity and ability to innovate and collaborate.

How do we reduce the threat state? Offering free yoga classes and mindfulness apps is a nice thing to do, but not if they are being offered in place of addressing the real issue—how we are treated at work. For more on what we can do, read on.

2. Stress—find the right balance between challenged but not overwhelmed

It's not that the brain doesn't need any stress—we do need some to get ourselves going and to be motivated, but it is a fine balance. To work at its best, the human brain needs to feel challenged and stretched but not overwhelmed—at the top of that inverted U (see Figure I.2). When we reach this sweet spot, it's a state often referred to as "flow" by psychologists. The inverted U of performance illustrates this well. Too little stress and we faff around and get little done, too much pressure and we are stressed out and can't think straight. In far too many organisations, there is a tendency to push people to the right of that inverted U, to ask people to take on more, to get more done, and do more with less' But there's a limit and the human brain cannot work at its best if we are constantly feeling overwhelmed or frazzled. It's not a clever nor effective way to lead people.

Good people managers are those who tune into what each team member needs. Some (those who have a tendency to be on the left-hand side of the inverted U) might need tight deadlines to get them moving. Other team members might be more easily tipped over the edge into overwhelm. These people might need to be helped to focus on what they can control. If people feel that they are working in a very uncertain world, we need to give them certainty where we can: Being clear and reliable in terms of how we work with them and how we communicate with them (eg., 'We don't yet have much certainty about the next 12 months, but I will communicate with you every Tuesday morning what I do and don't know, and will answer your questions as best as I can').

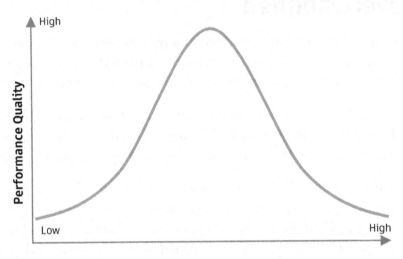

Figure I.2: Inverted U of performance.
There is an optimal level or arousal—leaders can ask themselves where they and their followers may be on this curve.

3. Subheader (H3) Pay more attention to the reward state and the brain's seeking system

Organisations now and in the future will need creativity, problem-solving, and collaboration. These are only going to be achieved if employees are in a reward state. It follows that leaders need to know how to activate the reward state in people's brains. When our brains are in a reward/toward state, the brain's seeking system is activated. We feel more engaged with work, are more curious, and can collaborate better. What helps to put the brain in a reward state? Many things. Feeling we are being treated fairly at work, having a positive narrative about ourselves, being treated with respect, feeling that our work is meaningful, having good relationships at work, being listened to, co-operating, being kind to others, having some choice, novelty, laughing together, feeling that we are mastering skills, and, on a less positive note, schadenfreude (particularly if we think it is a bad person who has got their just desserts).

The good news is that many small actions can help to create a work environment which activates the reward centres of the brain. For example, after a session on applied neuroscience, two leaders in an engineering firm decided to change their approach. They thought about threat and reward and the inverted U of performance and how they could apply these to the team. They became more open with information and talked more about a shared goal. They considered autonomy and made sure the group was clearer about what they could influence. They created an environment where people could openly share their thoughts. They set short-term goals and celebrated achievement. Over the following months, for the first time in years, they saw the cost-performance index begin to move. It improved by 70% in the first two months. "In what was a very difficult challenge in an environment that historically would have been highly tense and stressful, we now see a transitioned working group, working together and enjoying it while still delivering commitments to a strict timeline.... Focusing attention gives results... empathetic understanding and positive recognition of contribution has helped."

4. Purpose

To have the energy and drive constantly to change requires a sense of purpose. Purpose–that sense that our work is meaningful and makes a difference–has been covered in many books, articles and research papers, so I'll keep this short. Just to say that there is a big difference between being told that our work makes a difference and experiencing or seeing that impact for ourselves.[1] Research found that allowing doctors to spend one day a week on work that had most meaning for them reduced burnout.[2]

5. Social connection

We are deeply social creatures. We recognise this in our personal lives, that relationships matter, but historically this is an area that organisations have undervalued. We expect employees to walk through the workplace door or turn their laptop on in the morning and somehow relationships should not matter in quite the same way. But they absolutely do. Remote and flexible working bring many benefits, but one of the areas we are going to have to work hard on is maintaining good social connection when people are geographically scattered.

When we have a strong sense of social connection and belonging, we are more resilient and will keep going for longer at tougher tasks. In an era of constant change, this will be critical.

There are many research studies that illustrate the difference made by being in the presence of someone we like and trust. In one study[3], people were asked to estimate an incline—the brain tends to overestimate the steepness because it doesn't want you to expend energy unnecessarily; when people are in the presence of a friend, their estimation of the incline drops. Social support can influence our visual perception and our estimation of the difficulty of a task.

6. Change—moments of insight

So often during change, leaders will disappear into darkened rooms, often with consultants, for weeks and then emerge with the strategy or plan. They will then go into broadcast mode, telling employees why the changes are good for customers, for employees, and for all stakeholders. They then wonder why so many employees dig their heels in and don't want to make the changes. They have missed an important step. Leaders often forget that they have had the chance to read the data, to reflect, and then to have their moments of insight about why change is needed—that 'aha' moment: 'I get it—I can see why we need to do X or Y.' Having that moment of insight makes a big difference to our brains. We feel much more committed to an idea if we have reached it via insight rather than being told. We need to give employees that same chance to 'get it' that leaders have had.

7. Control, choice, and autonomy

Choice and a sense of autonomy are very important to the brain. Throughout a change we need to give employees a sense of some control. Back to that IT team member in the bank who was wondering why people want the latest smartphone but don't like new IT systems: One of the big differences here is choice. A change that we have chosen feels different from a change that is imposed upon us. Goals that we have chosen feel different from goals we have been given by others (and we are more committed to the former). So, the question for organisations in continuous change will be, 'where can we let go, and where can we give people a bit of choice or control?' It will make a significant difference to how they feel about change.

8. Belief that we can change

To change takes effort and energy. Our brains want to conserve energy and before we embark on change our brains have got to believe that it's worth it and that we are capable of change. Mindset matters. We all need a positive narrative about ourselves.

As one HR director said, "I decided to change my approach, and focus on the positives. This didn't mean ignoring issues, but it meant accentuating the positives in our performance conversations. I was delighted with the results. I saw a big lift in the self-belief and confidence of two of my team, and as a result got more drive, energy, and productivity from them both."

9. Equity and fair play

Our brains are very sensitive to fairness. Out on the Savannah, if we got our fair share of food we were much more likely to survive. When we are going through change in the organisation our need for a sense of fair play goes up: We have a sense that if there is going to be change around here, then I want to know I have as good a chance as the next person of getting the role I want.

10. Past change experience

One of the factors that has a significant impact on how we feel about change is our past experience of it. The brain is constantly comparing current circumstances with past experience. If we have been through change before and it was a positive experience, then we are more likely to take a positive view of upcoming change. If, on the other hand, our experience of change was painful and unpleasant, and more change is coming, then we are likely to see change as something we want to avoid. The more we can make change a good experience, the better this bodes for the future.

Six more things you can do with the ACCESS framework

If the future of work is about having the ability to change constantly and be flexible and adaptable throughout our working lives, then we need to know what is going to help our brains to do this. How do we look after our brains in a world of constant change?

Source: Scarlett, H. (2019). *Neuroscience for organizational change: and evidence-based practical guide to managing change* (2nd ed.). KoganPage. ISBN-13: 978-0749493189

Figure I.3: The ACCESS framework is a memorable summary of brain care for people.
Leaders can use the ACCESS framework on themselves or with their teams.

ACCESS framework

We know that a healthy body needs a certain amount of vegetables and fruits each day, but what does a healthy brain need? What do our brains need to be capable of dealing with constant change? The ACCESS framework (see Figure I.3) summarises six factors.

1. Acknowledge good things in your life

As we know, to reach its goal of survival the brain tends to home in on the negative and pay less attention to the positives in our lives. It is useful to learn to counter-balance this tendency. As we explored earlier, when we are in a reward state, we are more resilient, creative and open to new ideas. Each day, just thinking about three good things in our lives subtly shifts the brain's filters: The more we think about the good things, the more good things we are likely to see. At the end of each working day, we can also reflect on what we have achieved (these can be tiny tasks). The brain tends to ruminate on all the things we didn't do, so we need to remind ourselves of what we have achieved.

2. Calm the brain on demand

Emotional regulation is an important skill. In a world—both inside and outside of work—that is full of triggers for stress, we need to find a way to be able to calm our brains when necessary. Practising mindfulness (or meditation) has become more popular in western countries in the 21st century. There are plenty of neuroscience studies that illustrate the benefits it brings. Using the breath is another useful technique. Even something as simple as taking a deep breath in and slowly out, and then pausing at the end of the outbreath, can help. Breathing slowly and pausing at the end of the outbreath activates the parasympathetic system (sometimes referred to as the 'rest and digest' system): This is one of the fastest ways to send a message from the body to the brain that you are calm: The brain thinks that if you have time to breathe that slowly and pause, you cannot be in fight or flight. Whatever technique works, it is important to find one that gives us the ability to calm the brain on demand.

3. Challenge your brain

Organisational change often means we have to learn new ways of doing things and our brains can find this hard (it's easier and less effortful for things to stay as they are). But challenging the brain and learning new things is good for us. Learning new things helps us build cognitive reserve and it's cognitively protective. Research into 'super agers'–people who have sharp brains into their late years—reveals that these are often people who have chosen to carry on learning and applying that learning throughout their lives. That phrase, you can't teach an old dog new tricks: Not true. As long as the old dog wants to learn new tricks, it can. And what's more, it's good for the brain.

4. Exercise

We know that exercise is good for the body and it's also good for the brain. It's probably one of the best things we can do. We are not designed to sit at a desk or bent over a laptop for many hours a day. Exercise not only gets blood to the brain but it also helps to generate brain-derived neurotrophic factor (BDNF). BDNF has been shown to help the brain create new brain cells, particularly in the hippocampus, a part of the brain important in memory. There are now many studies illustrating the link between exercise, especially exercise in nature, and cognitive benefits.

5. Social connection

We have touched on this. We are deeply social creatures and research suggests that throughout our lives having someone we can talk to is important to our resilience.

6. Sleep

People used to boast about how few hours' sleep they could get away with. Now it is recognised that the majority of us need seven-or-more hours of sleep. Our brains are very busy during sleep, learning and consolidating memories. Lack of sleep weakens our cognitive performance and our ability to regulate emotions.

Change is increasing faster than before. Our Savannah brains developed for a world that was slower in terms of change. We need to have some understanding of the brain so that we can work with this knowledge and equip organisations and the people who work in them to be adaptable. We need to reduce the threat state and create an environment where the brain's reward and seeking systems are activated. Then we will have workplaces where people can be problem-solvers, curious, creative, and collaborative.

References

[1] Grant, A. (2014). *Give and take.* Orion Publishing.

[2] Shanafelt, T.D., West, C.P., & Sloan, J.A. (2009). Career fit and burnout among academic faculty. *Archives of Internal Medicine, 169*(10), 990-995.

[3] Schnall, S., Harber, K.D., Stefannucci, J.K., & Proffitt, D.R. (2008). Social support and the perception of geographical slant. *Journal of Experimental Social Psychology, 44*(5), 1246-1255.

Further Reading

Arnsten, A.F.T. (2009). Stress signaling pathways that impair prefrontal cortex structure and function. *Nature Reviews Neuroscience, 10*(6), 410-422.

Yerkes, R.M., & Dodson, J.D. (1908). The relation of strength of stimulus to rapidity of habit formation. *Journal of Comparative Neurology and Psychology, 18*(5) 459-82.

CHAPTER 2

Mental health in workplaces and in organizational change

a dialogue between Newton Cheng and Paul Gibbons

Mental health is one of the most critical issues we face as a world. Despite being 1.5 times more prevalent than cancer, mental health conditions are worsened because of the surrounding stigma; they go undiscussed, undiagnosed, and untreated.

Leading businesses are beginning to address the issue, but we have a long way to go. One such frontier is organizational change. If you were to survey the "change management classics," you would not find a mention of mental health—despite the obvious links. This is an oversight, and while change consultants should not "play therapist," the time has come to understand better the reciprocal relationship between mental health and organizational change.

In this chapter, after a brief prelude, Paul Gibbons and Newton Cheng first share their personal stories, then discuss systemic issues, leadership and cultural aspects, and what those mean for organizational change. Newton is director of global health and performance at Google, a family man with two young daughters, and an athlete who competes in the World Powerlifting Championships.

Mental Health Facts and Background

Mental illness, according to the U.S. National Institute of Mental Health, afflicts one in five of us at any one time. **Half of us** will experience an episode during our lifetime. It is possible, perhaps likely, that because of the stigma associated with mental illness, these estimates are grotesquely too low—the problem may be far worse.

Nor do most people get help. The National Institute of Mental Health goes on to say that about half of all people (and more than half of all men) go untreated.

Success provides no cover. Walking on the moon didn't help Buzz Aldrin. Leading the United Kingdom through World War II didn't help Winston Churchill. Winning two Oscars didn't help Sir Anthony Hopkins.

Occasionally, mental illness makes the papers, usually tragically: Robin Williams, Kate Spade, and Anthony Bourdain ended their own lives prematurely in just the last few years. (see Figure II.1). It is tempting to reach for the iceberg metaphor, but this iceberg has a few dozen celebrities above the waterline and a billion sufferers below it.

FAMOUS DEPRESSION SUFFERERS

Figure II.1: Famous sufferers from depression.
Depression not only affects some celebrities, but also affects more than a billion people worldwide.

Mental health is a huge global issue, crucial for workplaces, yet largely ignored by change experts.

After sharing their personal stories, a few of the questions Paul and Newton discuss are:

- Workplaces have become increasingly involved. Have they done enough? What is next?
- How does stigma confound prevention and treatment?
- What are the leadership and culture issues?
- How should models and methods in organizational change adapt?
- How should mental health concerns affect how organizational change is led?

Mental health stories

Newton's story

PAUL: Newton, what sparked your coming out on mental health, becoming an activist, so to speak?

NEWTON: As an Asian American and as a man growing up in the Midwest, I was taught that you don't show weakness. You certainly don't cry at work. But there I was. We were kicking off a video meeting with a check-in question: "How are you doing?" Not in a back-slappy "how ya doin' bro" manner—to which the only acceptable reply is a version of "living the dream." That day, I felt compelled to say something real.

I started to cry and said, "Right now I'm struggling because the number of days that I'm proud of how I'm showing up as a father is going down, and I don't know how to run that around." The truth, expressed in tears, was that I was not doing well. The mask dropped. For years, I had been shouldering my responsibilities as a father, academic high-achiever, executive, and powerlifter: Soldiering on, since my mid-teens, muscling my way, if you will, through a morass of anxiety and depression. The dam had burst (see Figure II.2).

What followed from my team at Google was an outpouring of support. They got it. Several had felt the same way. My co-workers were terrifically supportive, particularly my boss, who made it clear that they were personally and corporately there for me.

However, the shame of "losing it" did not disappear.

NEWTON WAS WINNING MEDALS, INTERNALLY, HE WAS SUFFERING

Figure II.2: Externally, Newton was winning medals; internally, he was suffering.
Mental health affects high performers. Visible success can mask inner distress.

I decided to put the mask back on and suck it up. The pandemic had struck. Google, like every business on the planet, had to adapt. My role involves supporting Google's people when they are struggling, so I took on even more responsibility, leaning in even further. By the end of 2020, it was getting harder to get out of bed. April 2021 was the first time there was a morning when I couldn't get out of bed because I was paralyzed by a sense of dread.

PAUL: I want to hear about your journey back. But first, I want to thank you for the lives you've touched and improved by sharing your story. I sometimes say, "Transform your wounds into a gift to the world." You've done that through your social posts, activism, and on the speaking circuit.

Paul's story

NEWTON: Thank you—that means a lot. You have a personal story too?

PAUL: I do. Curiously, while I've been very open about my 30-year drug and alcohol recovery, it puzzles me that I've let social stigma keep me silent on my other mental health issues. You've opened that door for me. So here we go… For the first time anywhere…

Like you, I've been dogged by depression since I was a teenager. Later, in my 30s, I self-diagnosed as an alcoholic and, at fifty, was handed a diagnosis of ADHD.

Though I was smart, starting to audit college classes at age eleven and earning ridiculous money as a trader in my early 20s, my cocktail (haha) of issues meant I often crashed and burned. My lows were very low—I wasn't capable of holding a job for six years and lived on couches until I got sober. Then came a meteoric climb through the consulting profession and founding my own firm. When that firm hit rocky times in the late 2000s, I had to sell—walking away from my startup, getting divorced, and moving countries brought me very low again.

The happy ending is that juggling my roles as a father, professor, poker pro, writer, and speaker have been terrific the last two decades. Then, in 2022 and 2023, I went back to corporate life and ran into a very toxic culture, which took its toll. 2024 is another one of my rebooting years.

NEWTON: I'm glad we finally met. I've been reading your books, you've been resharing my posts, and here we are.

Mental health recoveries

Newton's recovery

PAUL: You've amazed me by your depth of knowledge on leadership and personal development. I was like, "whoa, this dude is an engineer and powerlifter; how does he know about that?" You must read a ton. And, as I understand from your public posts, you put all the leadership and personal development work into your recovery and your work at Google.

What did your recovery look like? Give us the long version.

NEWTON: The pandemic had put enormous adaptive strains on Google and its workforce, just like so many others. Part of my role is helping people through such times. But the increased pressure of supporting Google's workforce began to take its toll, acutely in my case.

Freud said that love and work are the cornerstones of our humanness. I wasn't doing well at either.

However, with the incredible support of my manager, my team, and Google's culture at large, I went on short-term disability leave in Jan 2022 to care for my own mental health. Upon returning, the most common question that I was asked was, "What did you do on break that helped you recover?" Here are a few things.

During **week one**, honestly, I drank too much with a good friend. It genuinely helped me feel connection and belonging, but drinking is also a long-standing coping mechanism for me. I gave myself permission to "go there" but knew it wasn't the way forward.

PAUL: Like so much, it can be healthy and life-affirming in the right context. In the wrong context, not so much. It takes wisdom to know which.

NEWTON: During **week two**, I jumped into action. I applied my athlete's mindset and gathered all the books and experts on recovery, healing, resilience, and self-improvement around me. I pieced this into a logistically sound plan.

Then, I attacked the plan.

By the end of week two, I realized I had created another job for myself. And this wouldn't work. I had made a clear, logical error, trying to address my "overwork" by . . . aggressively working my way out of it.

PAUL: You can't address overwork with . . . more work. Duh. But human beings default to what they excel at and what they know rather than what might work better, rather than trying something different. You were in a tough spot, so you fell back on your super-achiever superpower.

NEWTON: In my heart, I knew that, but my old beliefs told me I could "figure it out" and "get it done faster." Which is how we ended up on mental health leave in the first place.

By **week three**, I tried something different.

I hung on to some aspects of week two's "drive for results"— meditation, exercise, and regular reflection. However, I added (or subtracted) some things. I subtracted many of the "go-getter" activities and substituted more:

- Space for stillness,
- Blocks of time with nothing planned,

- Time wandering, and
- Unstructured reflection (versus reading and answering prompts).

In hindsight, I needed a lot of space to 100% "do nothing" which proved essential to my healing process.

During **week four**, I added two tools that were so simple yet so life changing.

The first was "Daily Pages" from the book *The Artists Way* by Julia Cameron. This exercise is used to get creative writers to stop self-editing as they write and instead just get things out on paper. For a set amount of time, you just keep writing without stopping to self-edit. My big "a-ha" was that when I write, I harshly judge my thoughts and words even before they hit the paper.

Writing those thoughts down was helpful input to the next (second) transformational input.

That second input was *The Work* by Byron Katie, which relies on four "liberating questions." For every negative thought, you ask yourself:

- Is it true?
- Can you absolutely know that it's true?
- How do you react when you believe that thought?
- Who would you be without the thought?

After that, you apply what Katie calls "the turnaround"—"a sentence expressing the opposite of what one believes."

Seeing those thoughts, first via *The Artists Way*, then repeatedly questioning them via *The Work* showed me they lacked actual power and validity.

PAUL: I still find it super cool that someone not in the leadership development business knows their way around some of the power tools as you do.

So, while knowing there are no silver bullets and that people's journey to recovery will be as distinctive as they are… What advice would you give to people?

NEWTON: First, people shouldn't take this as medical advice. Get medical advice from a professional.

With that in mind, I can't overstate the importance of seeing a therapist. When I went, I was thinking, "Maybe I'm a little depressed—just a little."

He blew me away by saying, "You are showing **major** signs of depression and anxiety." It wasn't like I was on the border. Still in defensive mode, I asked, "Isn't this just what working hard feels like? I've felt this way off and on since high school—and I've just powered through it." Having always felt that way, I had come to experience it as normal. He assured me it wasn't.

After that, I felt more able to talk to friends. Before that, I felt like I was uniquely flawed, just the biggest f-up in the world. But after talking to them, I saw, OK, maybe I'm not alone.

And then, with considerable trepidation, I talked to my team, peers, and boss at Google. And they stepped up big time.

But my private conversations with a professional empowered me to take the next healing steps.

Why don't you jump in and share your recovery story, Paul?

Paul's recovery

PAUL: There are two things I've done right and one that I'm still grappling with.

One thing I've done abundantly right is getting sober at age 32 and staying sober for 30 years. If you are an alcoholic and continue to drink, it doesn't matter how much therapy you get or medication you take; alcoholism will confound your healing from other mental health issues.

To fully recover, and I consider myself recovered after 30 years, you have to work on the root causes of addiction. During those early years in sobriety, besides the love and support from the Alcoholics Anonymous folks, I made personal development and recovery the mainstay of my life, including traditional stuff like therapy and getting trained as a Gestalt counselor, and some non-traditional stuff that bordered on the whacky sometimes. (Think of the whackiest Californian self-help workshop—guaranteed I've done it.)

That recovery path led me into leadership development and change management because leadership development and personal development are interrelated—synonymous isn't too much of an overstatement. Then, I got a gazillion coaching certifications and brought the tools I had used in my own recovery to leaders—as a coach and running a leadership firm. We helped leaders do the deep personal work needed to realize their full potential.

Then, I stopped personal development on a dime, which is the thing I'm grappling with.

But first.

The other thing (and this might qualify as wisdom, so listen up, folks) is the mindset I bring to episodes of depression.

I used to give myself an incredibly hard time when I got depressed and berate myself for all-nighters playing Civilization. But a therapist said, "That sounds kinda healthy. You push yourself a lot, you are a hyper-achiever, and a rubber band can't be indefinitely stretched without breaking. Maybe there are better things than gaming, but it is something you enjoy that helps you recover."

I cycle. Mostly, I'm an upbeat, productive guy. But sometimes I have crappy days, maybe even crappy weeks. Even when life is going amazingly well, I get these periods of depression.

When I get down, the thing I 100% emphasize most is not to berate myself, catastrophize, or allow my thoughts to tunnel into even darker places.

In other words, I don't get depressed about being depressed. Or anxious about being depressed. I radically accept that I might feel shit all day and not feel like doing much, and if I need to sleep, or workout, binge out on gaming, or watch TV, I allow myself to do just that. For a day or two. And, inshallah, it passes.

> *Editorial comment*: Newton reframed his negative thoughts—similar to CBT approaches; whereas, Paul approached his with acceptance and self-compassion—a Gestalt or Buddhist approach.

That helped me **a ton**—the whole "be kind to yourself" thing I learned from that therapist.

And it is really healthy for someone hyper-competitive like me to have areas I'm not competing in—because in all my hobbies, I compete super hard, like you, at an international level, but poker and bridge are my "sports" because I'm a nerd. Gaming, golf, and tennis I do just for fun, and I suck, but I don't care. And it is good to have those outlets.

NEWTON: It sounds like you focus on being as kind to yourself as you can be. And give the maniacally driven dude a break.

What is the thing you are grappling with?

PAUL: About 2004, when I was running my own business, I stepped 100% away from the personal growth and recovery that I had dedicated the previous decade to. I pivoted from reading a book a week on change, psychology, and spirituality to reading hard science and philosophy. People wonder why, as an atheist, I have two hundred books on religion and spirituality. Spirituality and personal development were my life's work, my purpose, my academic interest, and my profession. And then, I stopped.

The shift away from personal development toward hard science feels authentic and makes me more like the kid I was as a teenager, but I fear I threw the baby out with the bathwater. In 2024, I need to get back in the game, reincorporating some of that personal development and working on myself, so to speak, but truthfully, since we are being real here, I'm not sure where to start. It is hard getting back in—you go to yoga and are stiff as a brick; you try to meditate, and your mind is like sitting trackside at Formula One. And I'm not surrounded by a personal growth community as I once was when I ran my own firm.

NEWTON: Got it. It sounds like a journey of integration—or at least making room for both.

Shall we talk a bit about workplaces and mental health and then move on to change and mental health?

PAUL: Yup, you first.

Workplaces: EAPs are not enough

NEWTON: In the most callous dollars and cents terms, according to the U.S. Centers for Disease Control (CDC), depression **alone** costs employers $25-50 billion a year. According to Mental Health America, these costs may surpass employer-sponsored healthcare costs! Then there are harder-to-quantify costs, like absenteeism and lost productivity. Workplaces also care about engagement, workplace climate, retention, creativity, and collaboration behaviors. These, too, are likely impaired by untreated mental illness.

We now know that 98% of mid-to-large-sized companies offer employee assistance programs (EAP). That is the good news. The **bad news is that utilization of such programs remains abysmally low—between 2-8%**. While necessary, EAP programs that offer support after the fact are an entirely insufficient part of the solution.

We may read about extreme cases, such as a breakdown, but the more prosaic, daily story that affects billions is people being less present, less joyful, and less active with their loved ones. How many kids, partners, and neighbors feel the brunt of workplace-induced issues? In my case, I began to worry about my ability to fulfill my roles as husband and father. And those relationships nurture and heal me.

PAUL: Yes, there is the systemic stuff again—relationships heal and nurture, but also are among the first things to be damaged by impairment.

NEWTON: Having said that, there is much more to do, it is safe to say that workplaces have come a long way.

Merely decades ago, a worker's mental health was their own affair—they were expected to suit up and show up. Times have changed—however, the legacy of that attitude persists in many managers' minds, to "leave your personal issues at the door."

Changing managerial and leadership attitudes and behavior is one of the biggest levers for change in the system.

PAUL: I want to ask you later the "wave a magic wand" question on leadership and culture, but let me underline the systemic bit, if I may.

Mental health affects how well people look after themselves, and how well someone looks after themselves affects their mental health. That circularity is apparent, but people forget. I know for myself that when I feel depressed, the first thing to go is exercise. Research shows exercise may even beat anti-depressants in a controlled trial.

We have more of this circular cause and effect elsewhere: Mental health issues **damage workplaces**, and workplaces may **injure mental health** through mechanisms such as toxic culture, stress, overwork, burnout, and a hostile climate.

Leaders with untreated mental health issues may cascade their issues throughout their teams and companies. I had one of those. It wasn't fun for our team. In my couple of years at IBM, no fewer than a dozen people opened up to me about their mental health struggles caused by what they experienced as a "culture of fear," as some put it.

This all raises a critical point. IBM is rightfully proud of its stance and its EAP policies. Still, those policies made zero difference to the environment that (in my opinion) was having such a terrible effect on people's mental health. And that tells me that **leadership and culture are what matters**, not whether we can support people who have been damaged through toxic workplace situations.

Most firms today are good at awareness-raising and after-the-fact support. That isn't enough.

Subtle trap of awareness building

NEWTON: No lesser scholar than Dwayne, the Rock, Johnson said, "With depression, one of the most important things to realize is that you are not alone." The social and cultural aspects matter tremendously.

But let me give you a hot take on awareness raising.

Awareness raising contains a subtle trap. While awareness is necessary for action, it is insufficient on its own. Awareness without behavior change is a booby prize. For example, sometimes raising awareness of DEIJB issues may persuade leaders that the "box is checked." But being aware, abstractly, that exclusion is a real thing may persuade people that they are inclusive. That is BS. Like being aware that there is this thing called racism means you aren't a racist.

So, it is in mental health. Making people think they're less susceptible to mental illness because they're aware of it is the trap.

If the data that fewer than half of employees avail themselves of these options are correct, there is more to be done. Not only that, but by the time half of people seek support, there has been unnecessary suffering and damage to work performance and culture.

So, I think leading organizations have to look beyond what has become standard: Mental health services, free therapy sessions, and EAPs.

Kaleidoscope of systemic issues

PAUL: You are saying that treatment is only part of the picture, and the medical profession is only part of the solution. The problems and solutions are kaleidoscopic, with hundreds of variables, cause and effect blurred, and non-linearities and vicious cycles, as I described earlier.

How is Google addressing these super complex systemic issues?

NEWTON: We have worked a lot on burnout at Google, often referencing the great work of Christina Maslach from my *alma mater*. She defines burnout as "the outcome of chronic, work-related stress." Some researchers feel that burnout is merely a less stigmatized description of depression because the symptoms of burnout—such as exhaustion, cynicism, and lack of professional efficacy—overlap. Certainly, we can talk about burnout more easily than using the "D" word.

I'm lucky to work for Google, a firm that strives to be among the best companies in the world in how it supports the well-being, including mental health, of its nearly 175,000 employees. We are thinking hard about prevention and overall well-being so Googlers can thrive. As well as top-down initiatives, we encourage grass-roots efforts that include an employee-run mental health conference, at which a VP openly shared their own mental health story. That inspirational story helped and inspired me when I decided to speak up.

Another way Google catered to their employee' well-being during the pandemic was their reset days, or mental health days, also known as Google "global days off." This was their method of giving their employees the chance to recharge during the pandemic to take care of their mental health, switch off from work, and prevent burnout.

The systemic question is whether the modern organization, its structures, processes, policies, rules, culture, leadership practices, and strategy are aligned or in conflict with the mental health of their workers—a huge challenge, but one that many leaders are confronting, inside Google and more widely.

Mental health stigma

PAUL: I know another thing Google is taking more seriously is the stigmatization of mental health. Say some more about that?

NEWTON: Look around your next meeting. If there are ten people, statistically, two may be silently struggling as I was. They are silent because of the stigma. Through my "coming out," I had marginally increased everyone's ability to self-express, but that wasn't enough.

Google strives to lessen the cultural stigma around discussing mental health, knowing that stigma both conceals problems and worsens them, making people feel distant and excluded. We recognize the importance of a systemic approach, including facilitation of connection and belonging and creating a nurturing, built environment.

Addressing stigma depends on people collectively breaking an unspoken cultural norm of silence. Stigma is real. However, it's not separate from us. We have to choose to perpetuate it by explicitly silencing ourselves and implicitly silencing others.

But we are making progress in my view. There are two paradigms for addressing mental health. The **medical model** focuses on individual diagnosis and treatment. The second, newer, more systemic paradigm includes the importance of communities, culture, and environment. In clumsy terms, no amount of psychotropic medicine or therapy can fully heal those whose circumstances are oppressive without addressing the circumstances themselves. This, of course, includes toxic workplace cultures.

> ***Editorial comment***: Google's Aristotle project, starting in 2012, began to research such links—the organizational climate factors, such as psychological safety, that determined wellness and team performance.

If I go to your integral change model, Paul, you see the opportunities to intervene, both in the short and long term, popping up all around our ecosystem.

And I guess maybe there isn't even a "there" to reach—just pushing back the frontiers of progress.

I know you've written a bit about that bigger context of capitalism and mental health.

PAUL: I have an 80% finished book called *Culture, Capitalism, and Sustainability*. When I started working full-time for Deloitte and IBM, I had to park the project. But I do think about such significant questions.

Social capital was once built in neighborhoods, communities, pastimes, churches, and families, but with the centrality of work in U.S. society, work has become an essential source of social capital. As the U.S. Surgeon General says, "Humans get through difficulties when we are together, not when we are alone, when we feel as if nobody has our back."

The fact of the matter seems to be that work is a source of social capital, and social capital is critical to mental health. We can either nurture social capital and "profit" from the support it gives people or not.

So, the big-big questions are, ethically, what is the responsibility of business leaders to staff who spend half their waking hours, or even more, at work? Is it as little as making sure the work environment is **not harmful** to mental health, which is hard enough? In humanistic terms, we are talking about the difference between human suffering and human flourishing. But how far does the employers' responsibility extend?

Does it extend as far as creating a workplace culture and climate that nourishes well-being and not just prevents acute injuries, such as from bullying and discrimination? Do workplaces have a preventative, positive role that goes beyond supporting after-the-fact treatment?

And I think there are cultural factors from outside work, from broader society, and its attitudes toward work and wellness. For example, in Silicon Valley and lots of other places, overwork is a status symbol. Burning the candle at both ends, sucking it up, going the extra mile, taking one for the team, and similar are cast as laudable.

And they can be, but not unconditionally so. Sometimes individuals may not accurately gauge when going the extra mile is one mile too many. Some bosses may expect that mile even when greater team awareness would not demand it. Managers rightly praise and promote people who lean in, but a great manager must develop sensitivity to those who have taken on too much in their striving to perform.

Enough about capitalism. Here is the magic wand question. In an Economist survey of ten rich countries, 69% of workers said their boss influenced their mental health as much as their spouse did!

What mindset or skills would you impart to leaders if you could wave a magic wand?

Leadership and mental health

NEWTON: One thing I've seen in my travels and discussions around the country is that there is a seductive tendency for leaders to view mental health issues as an HR problem, not something they need to become comfortable or skilled at dealing with. The next generation of leaders will need to aim to develop leaders who can walk a tightrope between setting high standards and driving performance while nurturing pro-well-being factors such as rest, work-life balance, and exercise. Leaders also have a critical role in piercing stigma and in breaking through the silence that perpetuates mental health problems.

And I have a personal "hot take" on mental health in the workplace. The one thing I see missing most (that people only say behind closed doors) is that they crave a place to have authentic conversations about their struggles and how the system around them reduces and exacerbates them.

Two things are needed. Vulnerably showing your wounds and naming the hard feelings in front of your colleagues is incredibly hard. It requires courage. Two of our male vice presidents courageously did this—an inspiration to me.

Name it, even if it's icky, such as shame, guilt, or deep anxiety. This allows people to say, "I feel it too, or I felt it too." This validation is part of the "social cure."

PAUL: I could not agree more, but how many leaders have those skills? How many are prepared to be vulnerable? How many are prepared to take their foot off the gas and sacrifice short-term results? And how many are aware of the longer-term effects if they don't take their foot off the gas? That harkens back to the capitalist imperative, the need to grind out profit, and short-term thinking means people's well-being may be sacrificed.

I want to share two recent occasions in which I think, in retrospect, that I failed as a leader.

On the first occasion, I had a boss that went super toxic on me. It affected me so badly that I didn't act compassionately—I focused on what an a-hole they were being and not being compassionate—"I wonder if they are struggling right now?" This would have been very hard to do, to be compassionate when I was under attack. But after a year of reflection, I now see I could have asked whether they were hurting. It is hard, when under stress, to focus on "how can I help this person?" rather than "what a &*^-head."

On the second occasion, a colleague didn't seem to be engaged, wasn't making progress, was missing deadlines. Again, under the pressure of finishing our projects, I kept pushing harder. I should have, given my own history, been able to say "hey, are things alright with you?"

The problem with doing that when a partner (or perhaps a more junior follower) isn't performing is it can sound accusatory, not compassionate. The leader needs to switch from "how can I get this stuff done, quicker, slicker, better?" to compassion and care, asking with love—"How are you really doing?"

And it turns out, after a few months, my colleague did say, "Hey, I need a break, all isn't right over here." But had I been more sensitive earlier, they might have got help earlier (and saved myself a fair amount of angst).

I would like to think that if situations like this come up again, I can do better.

Some bosses are even explicit about the need to work harder. Jack Ma and his 9-9-6 system is 9 a.m. to 9 p.m., six days a week. Musk makes such views clear. Niraj Shah at Wayfair said, "Working long hours, being responsive, blending work and life, is not anything to shy away from."

And while I don't share their values, I appreciate their honesty. My early days in consulting were close to 9-9-6 some weeks, but it was never explicit—just an implicit expectation. That is much sneakier and harder to confront than a spelled-out policy. Those leaders, in a clumsy way, draw attention to the contradictions presented by the system; the conflict between a growth-profit-and-efficiency imperative of capitalism and the human, perhaps, balance, relationships, families, or wellness.

The problem is that those leaders are greatly admired for the most part. And their attitudes and standards may be copied far outside their own companies. I don't like that.

They put 100% of the responsibility on the worker without questioning whether their culture might be harmful. And while I would not legislate, as they do in Europe, maximum working hours, I believe that as a culture, we need to shift the corporate paradigm and have expectations and norms that consider the importance of mental health. I mean, is there anything more important?

What is your take on resilience? How is resilience part of the preventative mix?

Resilience

NEWTON: Some of our work at Google, at least with my group, is about building resilience, which the American Psychological Association (APA) defines as "the process of adapting well in the face of adversity, trauma, tragedy, threats, or even significant sources of stress."

Developing individual resilience is a laudable aim that fits Google's culture and is more broadly applicable than just at work, it supports people facing life's difficulties as well.

Some gurus on mental health promote resilience as a magic bullet. It isn't that. Individual resilience initiatives also provide a valuable set of tools for Google: Recovery, movement, mindset, and nutrition. This isn't about making Googlers as tough as Navy Seals but about getting the basics right. Are you moving, hydrating, and making good food choices? Are you disconnecting and unwinding? Are you nurturing your most important relationships? Are you getting enough movement? Are you eating enough plants and vegetables and drinking enough water? Are you not overeating pizza? How are your sleep practices?

PAUL: There is one problem with the idea of resilience—it **falls upon the worker** to tough it out, bounce back, ride the storm, or whatever metaphor applies. It takes the focus off the system and places the onus chiefly on the worker.

While essential to recognize that some people may fare better than others, **individual resilience** is just part of a preventative puzzle.

The most interesting research frontier is the relationship between personal attributes and individual responses to stressors, noting that some individuals may not experience adverse effects in specific situations, whereas others may be severely impaired. Contemporary research attempts to understand which biological, genetic, cognitive, social, emotional, skill, and spiritual factors create or result in an individual's resilience.

This environment-individual symbiosis presents a complex but fascinating research frontier.

You will be pleased to know that next year, I'm developing a resilience diagnostic questionnaire that will be app-based. I'm going to integrate the cognitive, emotional, social, physical, and neurobiological aspects so people can assess their resilience for themselves.

But hey, we've been talking about change to mental health systems, culture, and leadership—what about how organizational change needs to change? How does change management need to evolve?

What about organizational change?

NEWTON: You know the change literature better than me, so I could be missing something, but my sense is that change people have been tiptoeing around mental health—using words such as burnout, change fatigue, and stress. But let's name the baby—badly managed organizational change affects people's mental health. And I think ignoring the fact that people might be affected by trauma or other issues before we attempt a change is essential, and I believe during change, we need to be more explicit and watchful about its affect on mental health.

Change experts at Google and elsewhere, to make informed choices about change strategies, need to know which of those may lessen the impact of change on mental health and, likewise, which might worsen it. They need to be able to advise leaders on which kinds of change might be the most stressful.

PAUL: You crack me up. First, you say you aren't an expert, then you stick the landing perfectly. I could not agree more. Especially that we have been talking while trying not to talk about it—perhaps because of stigmatization?

When I began to research this topic, in typical fashion, I hit the books and academic literature hard. What is out there?

While a teeny-tiny bit is written about trauma and change, it is super narrow—a million miles from a systematic take. There is also a weensy bit of research suggesting "constant change" leads to increased sick time, turnover rates, and decreased productivity and satisfaction. However, I'm unimpressed. Using "constant change" as the independent variable seems daft. Which organization isn't in constant change?

Moreover, such self-report survey research obscures critical variables, such as how well the change is led or managed and what kind of change. To lump all organizational change, regardless of type, extent, or how well managed under a single research variable called "change" is to conceal everything interesting about the relationship between mental health and organizational change.

It is not too strong to say that there is just about nothing about mental health in any of the books, guru websites, big-firm methodologies, academic papers, or even random blogs about organizational change.

Nothing on mental health is "baked into" turnkey change management solutions developed decades ago—and understandably—we just didn't talk about mental health in the 1990s when those methodologies were cooked up, nor in earlier years from which research was sourced (e.g., Lewin, Kübler-Ross). Those models haven't changed at all, but that is a rant for another day. People might counter that depression is one stage on the change curve, so it does feature, but the change curve as a whole is nonsense. Even if it were valid, just because depression is a proposed stage doesn't equip us to do anything about it.

NEWTON: Given the importance of mental health to work and the ubiquity of change, the organizational change canon should include research on mental health. You are saying it doesn't?

PAUL: It does not, and it seems absurd. As with work and mental health, the relationship between organizational change and mental health is reciprocal. **Change, particularly poorly managed change, may introduce new stressors that impair mental health. And the "reservoir" of mental wellness will affect leaders' ability to deliver change.**

NEWTON: So how do we begin to fill that void? You've researched mental health and are leading the charge in the future of organizational change. What do you see?

PAUL: The future of organizational change is at the center of my purpose at work now. I want to puke every time I see some model from the 1940s or 1950s that is crap, and was way back then, being trotted out as today's wisdom. Or some organizational change poll on LinkedIn that has four wrong answers. OK, rant over.

Mental health is a critical part of organizational change. It is time, I feel, to explicitly include our knowledge of mental health, from promotion and prevention to treatment, recovery, and maintenance, in how we theorize about change.

For starters, since depression is so significant to the mental health picture, I think change professionals should know their way around the PHQ-9 (see Figure II.3), which is a brief questionnaire in everyday language that indicates whether or not someone may be depressed.

PATIENT HEALTH QUESTIONNAIRE (PHQ-9)

Over the last 2 weeks, how often have you been bothered by any of the following problems? (use "✓" to indicate your answer)	Not at all	Several days	More than half the days	Nearly every day
1. Little interest or pleasure in doing things	0	1	2	3
2. Feeling down, depressed, or hopeless	0	1	2	3
3. Trouble falling or staying asleep, or sleeping too much	0	1	2	3
4. Feeling tired or having little energy	0	1	2	3
5. Poor appetite or overeating	0	1	2	3
6. Feeling bad about yourself or that you are a failure or have let yourself or your family down	0	1	2	3
7. Trouble concentrating on things, such as reading the newspaper or watching television	0	1	2	3
8. Moving or speaking so slowly that other people could have noticed. Or the opposite being so figety or restless that you have been moving around a lot more than usual	0	1	2	3
9. Thoughts that you would be better off dead, or of hurting yourself	0	1	2	3

(Healthcare professional: For interpretation of TOTAL, please refer to accompanying scoring card).

add columns   +   +

Total

10. If you checked off any problems, how difficult have these problems made it for you to do your work, take care of things at home, or get along with other people?

☐ Not difficult at all
☐ Somewhat difficult
☐ Very difficult
☐ Extremely difficult

© Pfizer, with permission • Future of Change Management, Gibbons & Kennedy

Figure II.3: Patient health questionnaire (PHQ-9).
PHQ-9 is a way to recognize the symptoms of depression in yourself and perhaps others. It is less threatening than longer, older depression questionnaires.

Now, I'm NOT saying we ought to hand out depression surveys like confetti. But I worked as a coach for twenty years, and I met a ton of people who were suffering without getting help. We were in a coaching conversation, and I could ask, "How are you sleeping?" or, "Are you having trouble concentrating?" I put the PHQ-9 questions to work.

When they had symptoms, I didn't go into the therapist role, but I was able to name what I was hearing and refer them to a clinical psychologist or professional therapist.

And the gratitude and the relief they felt, knowing, as we said above, that they weren't the only f-ed up person in the room and that there might be a path to feeling better, was powerful. I've got people thanking me for a decade. One guy's wife came up to me at a swishy London party in tears of thanks for having helped her husband get support.

Now, coaching is only part of what change professionals do, but coaches, even though they don't explicitly deal with mental health, must be aware enough to point people in the right direction.

This comes with a warning. I mentioned this to a buddy in the change field, and they were horrified. But I think this needs to be on the table with the appropriate boundaries. And I believe strongly adverse responses may be part of the cultural stigma.

But let me ask you: Would it have been helpful to you if some of this language and some of these questions had been in the vocabulary of change professionals inside Google? Could you have received help earlier?

NEWTON: During Covid, for sure. Remote working affected some people exceptionally positively, and others negatively. Some of the adverse effects include loneliness, anxiety, and depression symptoms. Greater sensitivity around mental health, including

PHQ-9 questions and encouraging change leaders to check in more frequently and overtly, would have been valuable.

PAUL: But coaching is just one part of the change management professional's job, and it isn't always part of their skill set. We also spend time on change communication and on training, and more. I think **everyone needs to know the signs, not just coaches**.

I think we can build in some PHQ-9-type awareness into change surveys. I think we can build mental health awareness into workshops (and we do a ton of workshops in my game). I think we can build it into leadership training and user training where appropriate.

I think that if the change is stressful, as changes sometimes are, we can build EAP awareness into communication. We can specifically build resilience into leadership and management training.

I've summarized the themes we've covered in a Figure II.4.

MENTAL HEALTH IN CHANGE: SUGGESTIONS FOR CONSIDERATION

...change professionals	...managers & leaders
Learn to recognize physical and verbal signs that someone may not be doing "fine" (e.g., irascibility, apathy, cynicism.) Don't take "fine" for an answer when you are worried about how someone is doing.	
During highly stressful changes, for example, downsizing, be explicit about mental health risks and support.	Take care of yourself! (Put your oxygen mask on first.)
Make EAP and other support part of change communications, as appropriate.	Drop your own "fake it until you make it" façade. Model authenticity, vulnerability, and the willingness to ask for help.
Build PHQ-9 questions into change surveys, as appropriate.	Drop nonsense change ideas such as "sense of urgency" and instead have leaders promote the idea that change is constant - policies, technologies, markets, people, and processes are in continuous flux.
Ensure change training and communications respect diversity - how different groups and individuals respond to stressors such as the demand for longer hours.	Don't assume knowing about something, or being aware makes you skilled in dealing with it.
Include stress management, resilience, and mental health in training (for those affected by change).	De-stigmatize mental health - perhaps sharing personal struggles - make it "ok" to not be OK.
Create a space, perhaps support-group style, where people severely affected, even traumatized, can safely explore their feelings and heal.	Understand the importance of psychological safety on workplace outcomes, such as creativity, and vulnerable staff.
Be conscious of mixed messaging – we often want to help leaders execute - but we also want to teach them to press pause when the change toll is too significant.	Recognize how historical issues, e.g., bias, racism, exclusion, bullying, mean that different groups may react differently.

© 2024, Newton Cheng and Paul Gibbons • Future of Change Management, Gibbons & Kennedy

Figure II.4: Guidance for change professionals, managers, and leaders.
Change managers can alter some of their communications and training and, critically, coach and support managers to be mental health aware during their role.

So, I think, and I may get pilloried for this, that change professionals should use their role supporting change as a trojan horse to generate more mental health awareness.

NEWTON: Interesting. So, if change is sometimes a stressor, it sheds a harsh light on the system and may offer ways to influence the system. After all, we want agile, responsive, resilient organizations. We need to create cultures and supporting structures and leadership skills that support that.

PAUL: You want to wrap up for us?

NEWTON: Sure, I see incorporating mental health into how change professionals operate is a great idea. And I'm sensitive to your horrified friend, we need clear boundaries.

I'm interested in the leadership aspects most because since "coming out," I've been thrust into a global leadership role in the mental health conversation.

PAUL: You are kind of the Taylor Swift or Greta Thunberg of mental health?

NEWTON: Not quite. If only. Most importantly, leaders create spaces where people can get to know one another, share experiences, and support one another. That seems foremost for me in a conversation with many critical features. Having a community at work can make all the difference to how people feel.

What I also see is bringing together mental health people with the DEI and climate people. Calling out those other movements and exploring the intersectionality may give gravitas to this frontier.

I think everyone is part of the solution. What is one of the biggest fallacies in complex systems? Seeing ourselves and the system around us as separate, discrete entities. We affect and are affected by the mental health system. We need, particularly, to be aware of how the system silences us.

In summary, I was lucky, I feel, to have great leaders around me and a company that takes this seriously. However, a problem on this (global) scale is going to take all of us. There are still too few leaders who are prepared to be vulnerable because of perceived political risks; too few try to integrate best practices into their teams, and too few experiments for the sake of their teams.

And I think, optimistically, we need to get enough people engaged to get to a tipping point—and I think, with my work and yours, we are nudging toward that.

Collectively, as a world, there is too much lost human potential for us to ignore.

Further Reading

Gibbons, P., & Cheng, N. (2024, January). *Mental health and work (guest Newton Cheng)*. Think Bigger, Think Better podcast.

Gibbons, P., & Hollon, S. (2019, August). *Depression, how can you spot it? What can you do? (guest Steve Hollon)*. Think Bigger, Think Better podcast.

Lilienfeld, S., & Arkowitz, H. (2015). *Facts and fictions in mental health*. Wiley.

Maslach C., Leiter M.P. (2016). Understanding the burnout experience: recent research and its implications for psychiatry. *World Psychiatry, 15*(2), 103-11.

Muñoz R.F., Beardslee W.R., & Leykin, Y. (2012). Major depression can be prevented. *American Psychologist, 67*(4), 285-95.

Southwick, S. M., & Charney, D. S. (2012). *Resilience: the science of mastering life's greatest challenges*. Cambridge University Press.

CHAPTER 3

Neurodiversity that Excludes? Tailoring Change Initiatives for Neurodiverse Stakeholders

by Dr. Ignacio Etchebarne

Neurodiversity is a new frontier in building more inclusive organizations. In this chapter, Ignacio (Nacho) introduces an unconventional take on neurodiversity—and weaknesses in the current diagnostic paradigm with profound consequences for organizational change.

Not quite normal

We start with the tale of a friend, a former professor, who in mid-life sought a formal diagnosis to see whether he fit on the neurodivergent spectrum.

From a young age and through college, he had fallen asleep in school when insufficiently stimulated. Only when work was extremely challenging could he focus. His friends' jokes sometimes flew over his head. It seemed to take him three times longer to complete projects or grade a pile of student exams than others. Occasionally, everyday noises and physical contact were unbearable.

There were also positives. When focused, he could sustain focus for much longer than his peers. He hardly registered hunger, the passage of time, or fatigue. On teams, he impressed team members by finding unconventional, innovative solutions to complex issues. He had an insatiable curiosity and eagerness to learn and an obsessive pursuit of excellence.

These positives contrasted with his struggle to stay engaged in administrative, numerical, or routine tasks.

On balance, the negatives triggered self-criticism and shame. He hoped a formal diagnosis would help him find calm, self-compassion, and meaning. Over the years, before considering formal assessment, he had wondered whether he had attention-deficit/hyperactivity disorder (ADHD). However, his scores on questionnaires were below cutoff thresholds.

He was so-called normal. But he didn't feel normal. There was pain. He felt out of step with the world, even sometimes with his partner.

Our friend's profile isn't neurotypical. His challenges and "superpowers" are unique, his flaws difficult but tolerable. Nor would he normally be labeled neurodivergent because he falls outside diagnostic criteria for dyslexia, dysgraphia, dyscalculia, ADHD, autism spectrum disorder (ASD), clinical depression, and/or anxiety disorders. He has some of each, but scores "subclinically" according to the DSM-V (Diagnostic and Statistical Manual of Mental Disorders, 5th ed.).

These diagnostic criteria penetrate the societal view of conditions and inform how schools and workplaces classify people. (In schools, this affects whether a child with such challenges may receive extra support.) However, some neurologists, psychiatrists, and clinical psychologists wonder how to assist individuals with similarly eclectic profiles. Other researchers, such as Teresa Torralva's team, discovered people who display "high-functioning attention-deficit disorder", which is only identified through specific tests.

Progress is being made. The neurodivergent classification criteria are being broadened to include people such as our friend. This new paradigm doesn't see our friend as "a problem," they just have a unique brain! Diversity, equity, and inclusion (DEI) initiatives also include these people.

Problem solved? Not yet.

Neurodiversity and the future of work

Our friend is me. Through the challenges and growing pains, finding self-acceptance continues. I survived graduate school with a PhD in psychology, and my obsessive-compulsive attitude toward learning and self-development has brought many "goodies." Today the consulting firm I co-founded (Human Insight or HI) flourishes beyond my wildest expectations.

I've been lucky to be part of a colleague community who welcome and support my "weirdness." However, for the millions of neurodivergent types inside monolithic corporations—with narrowly specified recruitment and performance management criteria where managers may be uninformed or unsympathetic—the struggle is less easy.

Looking ahead to the future of work for these people, we should wonder whether the neurodiversity movement has sufficient answers. Below, I share my research on a topic understandably close to my heart. Moreover, my consulting firm has developed an open-source approach from which I hope readers can extract what's useful, adapt it to fit their needs, and take up a notch their approach to the neurodivergent.

Before we talk about solutions, we answer the question: What are the limitations, shortcomings, and contradictions within the traditional neurodiversity paradigm?

Limitations of the current paradigm

The neurodivergent movement enables greater opportunity for a significant number of individuals who were previously excluded from employment opportunities and/or stigmatized worldwide. It has raised awareness that "neurodiversity" involves both neurocognitive deficits and neurocognitive surpluses. Distinguishing between skill deficits and skill differences is crucial to unlock divergent potentials.

Furthermore, the movement fosters an awareness of changes needed in the work environment, as well as in hiring and personnel selection processes, to effectively include individuals with neuropsychiatric conditions.

These laudable achievements benefit not just neurodivergent people, but both workplaces and society in general.

Despite progress, there is more to be done. Below are eight shortcomings to overcome for a future of more neuro-inclusive work:

Diagnoses come too late

First, individuals who (like me) have subclinical neurodivergent characteristics that don't meet formal diagnostic criteria often have difficulties that go unnoticed to the untrained eye. Unattended, this frequently leads to development of anxiety disorders, depressive disorders, or other mental disorders; and only then do they receive a formal diagnosis and the associated support initiatives.

Burnout and other mental disorders are expensive and risky

Second, waiting for an individual to receive a diagnosis often means they endure consistent performance failures, burnout, and/or emotional breakdowns. Additionally, individuals tend to conceal their challenges or postpone seeking help due to stigma and shame attached to a neuropsychiatric disorder diagnosis. Attempting to integrate a person only after an emotional breakdown or burnout is not only costlier, but also less likely to succeed compared to addressing the situation proactively.

Subclinical individuals are excluded

ADHD only gained formal recognition in the 1980s and its diagnostic criteria first focused on severe impairment. One friend was told, surprisingly, by an expert psychiatrist in the 1990s that "most people with ADD and few talents find themselves in jail."

Healthcare professionals are sometimes maligned

The neurodivergent movement wrongly portrays clinical diagnoses as pathologizing of diversity, when in truth, they play a crucial role in the dissemination of accurate information and in the integration of neurodivergent individuals into society.

Discrimination may be exacerbated

Categorizing people into discrete groups (e.g., neurotypical vs. neurodiverse) polarizes individuals, intensifying in-group and out-group biases (i.e., us-versus-them thinking). This inadvertently reinforces the very stigmatization and exclusion of neurodivergent individuals it was designed to combat.

Broad labels mislead

Current neurodivergent movements apply sweeping labels to neurodivergent populations (e.g., ASD, ADHD), creating an illusion of homogeneity. The reality is that two individuals with the same mental disorder diagnosis can markedly differ in their vulnerabilities and strengths, contingent upon their mental conditions, physical health, and backgrounds. Moreover, pure profiles are exceedingly uncommon, the norm being co-morbid profiles (e.g., someone with ADHD, social anxiety disorder, and dyslexia). Thus, broad diagnostic labels add very little informative value. In fact, from a neurocognitive perspective, we can find heterogeneity both within neurodivergent and neurotypical populations. Consequently, what's truly needed is a precise "laser" focus.

Knowledge is not integrated with other fields

The broad-label paradigm overlooks the opportunity to integrate with concepts such as psychological safety, management of change, executive function, neuroscience, and behavioral science.

Badging

Diagnostic labels carry significant weight, and when used it's easy to fall into what's known as "badging" or labeling, where we lose sight of the person, reducing them to their diagnosis. This is a major challenge in the neurodiversity movement today (i.e., being able to see the person rather than the labels, or at the very least, seeing the person behind the labels). In Figure III.1 below, we can see an example and its wonderful reframe, created by Chris Arnold. It's not the same to say, "Greg Wilson, who has ADHD and dyslexia," as it is to say, "Greg Wilson, who has the ability to approach situations while keeping a broad perspective and possesses impressive creativity for problem solving…oh, and yes, he also has ADHD and dyslexia."

Figure III.1: Badging, disability, or different abilities?
Do labels and badges help? Focus on deficit labels conceal strengths.

Beyond the normal-not normal dichotomy

Figure III.2 illustrates how the medical profession has historically emphasized pathology, deficits, or dysfunctions over high performance when combating psychiatric and neurological illnesses. Individuals are non-clinical or "normal" when their usual ability to inhibit impulses, solve problems, and adapt to change meets a predetermined standard. Those whose brains hinder them from achieving this ideal or standard functionality are labeled as "clinical," with "executive deficits," or "dysexecutive."

NEUROPSYCHOLOGICAL AND PSYCHIATRIC PERSPECTIVE

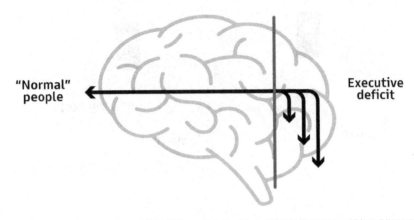

Figure III.2: Neuropsychiatric perspective of health and illness.
The psychiatric, or diagnostic paradigm of the medical profession classifies people bluntly, which overlooks performance strengths.

The neurodiversity paradigm depicted in Figure III.3 introduces a more balanced perspective on these phenomena. To its enormous credit, the movement proclaimed, "Beware! Deficits and differences are distinct things!", to refocus attention on the differences. People diagnosed with a neuropsychiatric disorder not only have deficits, but also often possess supra-normal capabilities (i.e., neurocognitive surplus). This realization prompted a reclassification of so-called normal individuals as "neurotypical," differentiating them from "neurodivergent" individuals with atypical brain functionality.

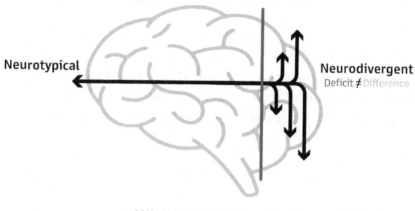

Figure III.3: Neurodiversity movement perspective.
Neurodiversity contains variation where difference does not always equate to a deficit. Who gets to decide which is which?

This is a pivotal concept that the neurodiversity movement introduced: Difference doesn't equate to deficit, or better said, diffability doesn't equal disability.

> ***Editorial comment***: In other words, the temptation to cleave the world into normal and not-normal, neurotypical and neurodiverse, is one to be resisted. As Darwin said, life has "endless forms most beautiful and most wonderful."

Focusing on deficits excludes people with extraordinary talents, it sidelines them and denies them opportunities. The neurodivergent movement shows us that along with formal diagnoses, sometimes were found great strengths. This new perspective is expanding, having been picked up by the media. In late 2022, Netflix released a TV series called *The Play List*, which is the story of Spotify. During episode five, a founder of PayPal self-identifies as being on the autism spectrum and diagnoses the co-founder of Spotify with ADHD. They boldly claim (without much evidence) that "the average rate of neurodiversity is 5%, whereas in the tech industry, it's 30%."

Researcher Nancy Doyle illustrates examples of these neurodivergent strengths in her book *Genius Within* (see Figure III.4). For example, individuals with ADHD sometimes have higher creativity, hyper-focus, energy, and passion than neurotypical people.

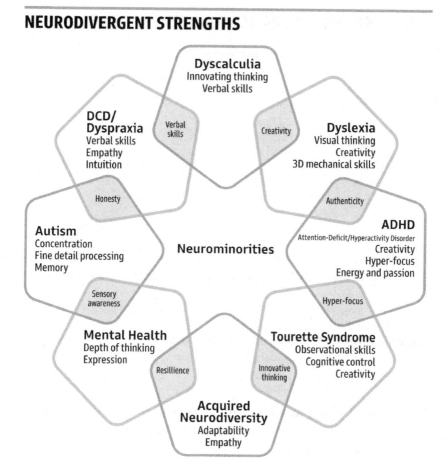

Figure III.4: Neurodivergent strengths.
People with neurodiverse labels have underrecognized strengths.

In Figure III.5 Nancy Doyle and her team compare the cognitive diversity of individuals with ADHD, dyslexia, and developmental coordination disorder (DCD, formerly known as dyspraxia) with the profile of neurotypical individuals.

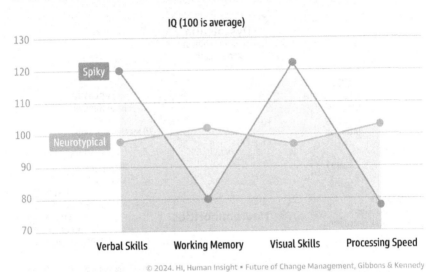

Figure III.5: Specialist vs. generalist cognitive profiles.
Again, neurodivergent individuals comprise deficits and strengths despite a traditional focus on deficits alone.

According to Doyle and her team's research, neurodivergent individuals exhibit significant intrapersonal variability in their cognitive skills (greater than two standard deviations), forming a more pointed or specialist cognitive profile. On the other hand, neurotypical individuals present a flatter (between one-and-two standard deviations) or more generalist profile. However, both groups have peaks and valleys, some more pronounced and others subtler. This presents us with a huge opportunity to give more visibility to variation to continue our development. Not all of us fall within the autism spectrum or obsessive-compulsive spectrum—because not all of us have a diagnosable disorder—yet we all lie on the neurodiversity spectrum. It's crucial not to confuse the former with the latter. Of course, this last statement should not be confused with the ignorant, invalidating, and denialist practice of dismissing neurodivergent diagnoses with phrases like "we're all a little autistic," "we're all on the [autism] spectrum," or "autism is so overdiagnosed these days," and so on.

Doyle also cites a challenge for neurodiversity to "incorporate both the high-performance potential of specialists and the competence of generalists." From my perspective, that's only one part of the challenge. The other part is how to identify deficits and strengths in each person (neurodivergent or not), reducing the former and enhancing the latter.

Neurodiversity of so-called neurotypicals

No person develops fully 100% of their brain's capacities. We have different strengths and vulnerabilities, and we all can improve. Nor can neurotypical individuals be lumped together.

Even in the neurotypical population, you can find neurodivergent characteristics in terms of strengths and deficits. For example, while developing the Y-BOCS scale for treatment personalization of people with obsessive-compulsive disorder (OCD), researchers have found that many "normal" people (or at least those without OCD) also have some compulsions and obsessions. However, that doesn't mean they have OCD. The diagnosis of OCD only applies when the frequency, persistence, and distress caused by obsessions and compulsions are so significant that they hinder a person from having a good quality of life and achieving their goals. Only in that scenario can we correctly state that the person has a disorder. This is where the person falls on the right side of **Figure III.6**, displaying a more visible neurodivergence.

NEURODIVERSITY SPECTRUM

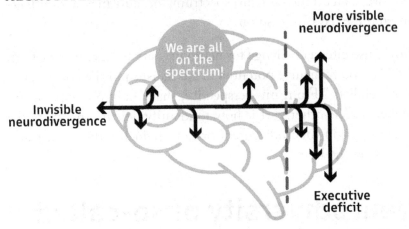

Figure III.6: Spectrum of neurodiversity.
Neurodiversity can be visible or invisible and invisible neurodiversity is equally important as visible.

This also applies to other mental disorders, such as generalized anxiety disorder, which is characterized by chronic and persistent worry. But who hasn't experienced intense worries at some point in their life? When does worry become clinical?

The failure to acknowledge this continuum between neurodiverse and neurotypical profiles is the root cause why countless people are left behind by the system, due current categorizations such as having "subclinical" symptoms, with "shadow" conditions, "intelligent but disorganized," "worried well," or "self-saboteurs." In your teams, among your peers, and in your family, there are individuals with these characteristics. You probably don't realize the daily cost because they've found ways to hide it by leveraging their strengths (e.g., high IQ) to compensate for their subclinical deficits. For instance, in my case, I compensate for my ADHD symptoms with my obsessive personality traits, such as perfectionism. It's not always obvious, but it is present.

Human Insights' awareness toolkit

In a complex and evolving area such as neurodiversity, there are no easy answers or panaceas. Still, my consulting firm—co-led with María Laura Buscaglia, Magdalena Oneto Gaona, and Clara Olivera—has tailored its leadership development tools with neurodivergent, behavioral science, and neuroscience in mind.

Here we share but two:

- Understanding the importance of executive functions for both neurodiverse and neurotypical populations, and
- Our leadership development framework with four neurodivergent-aware strategies.

Executive brain functions and neurodiversity

The executive functions of the brain are a group of higher-order cognitive skills that direct and coordinate other brain functions. These processes are key for successfully managing complex and uncertain situations. They allow us to regulate our emotions, resist temptation, maintain focus, pause our mental autopilot, take time to think before acting, mentally play with ideas, and solve novel and unexpected challenges. When using these skills, the area of the brain located just behind our forehead, known as the prefrontal cortex (PFC) activates (see Figure III.7).

These areas are essential for learning from experience, imagining, and creating our desired future.

There are several categorizations of executive functions. The simplest has three dimensions:

- **Inhibitory control,** through which we regulate our impulses (our mental brakes and accelerators).
- **Working memory,** which is our ability to hold multiple ideas in our consciousness simultaneously, such as when analyzing a complex situation.
- **Cognitive flexibility,** or our ability to let go of past action plans or interpretations and adapt based on changes in our context or access to new information.

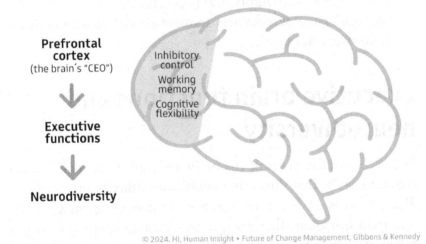

Figure III.7: Executive functions of the brain.
Imaging research localizes executive-level functions in the prefrontal cortex in humans, but this finding can mask the diversity of capability.

Well-developed executive functions explain a great deal of success in life, but no individual "scores" perfectly on all dimensions all of the time.

At HI, we find a three-dimensional executive-function classification too broad to offer practical use. We instead use a five-dimension executive-function classification, or STEER assessment, with the following categories:

1. **Self-observation and empathy**: This dimension evaluates metacognition and cognitive flexibility.
2. **Time management and organization**: This dimension evaluates time management, organization, and working memory.
3. **Emotion regulation**: This dimension evaluates emotional control or regulation.
4. **Effective planning**: This dimension evaluates goal-directed behavior, prioritization, planning, task initiation, and behavioral flexibility.
5. **Regulation of behavior**: This dimension evaluates behavioral inhibition and sustained attention.

Executive-function self-assessment

I invite you to uncover your own subtle neurodiversity or reconfirm your visible neurodiversity, and gain awareness of your neurocognitive profile by completing HI's Neurodiversity Spectrum Assessment (Hi-NSA®). We use this tool in leadership development and team management to become a "STEERing" force of action (see Figure III.8).

While self-report instruments are one of the most widely used resources globally for talent development, they all share the same limitations and vulnerabilities. They rely on the accuracy of self-awareness of those who complete them. In other words, these instruments do not evaluate how people really are **but rather how people think they are**. Therefore, the reliability of their results is only as high as the reliability of their judgments. In this sense, the proposed self-assessment of executive skills shares this same limitation. Nonetheless, they serve as an excellent catalyst for reflection to drive personal development and change management forward.

By identifying your executive profile, you can learn to use your executive strengths to compensate for your vulnerabilities and enhance your performance in work and interpersonal contexts. Additionally, by identifying the executive profiles of others, you can better appreciate their strengths and vulnerabilities, empathize more effectively, and discover how to leverage each other's complementary executive skills.

NEURODIVERSITY SPECTRUM ASSESSMENT (HI-NSA®)

As you answer, consider the following: **How is your performance in the following executive skills today?** Try to use the whole scale and be as honest as you can.

1. Self-observation and empathy:
Ability to self-monitor, register the consequences of actions, and consider multiple perspectives.

0 1 2 3 4 5 6 7 8 9 10

IF IT'S A VULNERABILITY...
You hear others say, "I can't believe you didn't notice that!" You struggle to understand others' reactions.

IF IT'S A STRENGTH...
You tend to view situations from multiple perspectives and reflect afterward on the impact of your actions. Others feel understood by you.

2. Time management and organization:
Ability to estimate and allocate time to meet agreed deadlines, and the ability to create and maintain monitoring systems for effective tracking of information, materials, and actions.

0 1 2 3 4 5 6 7 8 9 10

IF IT'S A VULNERABILITY...
You notice clutter piling up quickly (unread mail, unsaved files, emails needing decisions).

IF IT'S A STRENGTH...
You enjoy the feeling of order, having everything under control at home or work.

3. Emotion regulation:
Ability to stay calm in emergencies, express emotions effectively, and achieve goals, regardless of how you feel.

0 1 2 3 4 5 6 7 8 9 10

IF IT'S A VULNERABILITY...
You're accused of being "hot-headed" or "alarmist." You notice your emotions spiraling out of control while others around you remain composed.

IF IT'S A STRENGTH...
You're someone others turn to in emergencies because they know you'll stay calm and make good decisions.

4. Effective planning:
Ability to identify top priorities, create an action plan, and adapt it quickly to changes in your environment, unexpected obstacles, new information, or errors (adaptability).

0 1 2 3 4 5 6 7 8 9 10

IF IT'S A VULNERABILITY...
You struggle to identify what's most important and to take action. You get frustrated and resist adjusting your action plan.

IF IT'S A STRENGTH...
You find it easy to prioritize, distinguishing the essential from the incidental. You feel you can easily adapt. You enjoy experimenting and learning through trial and error.

5. Regulation of behavior:
Ability to exercise self-control, act according to plan, think before speaking or acting, and maintain focus on medium- or long-term goals.

0 1 2 3 4 5 6 7 8 9 10

IF IT'S A VULNERABILITY...
You often speak or act without weighing the consequences. You easily get distracted from your goals or postpone excessively the execution of your action-plans.

IF IT'S A STRENGTH...
Before deciding, you tend to analyze your options and the impact of different alternatives. You start on time and stay focused on your action plan until completion.

© 2024, HI, Human Insight • Future of Change Management, Gibbons & Kennedy

Figure III.8: Neurodiversity Spectrum Assessment (Hi-NSA®) for STEERing leaders and teams.
Build self-awareness of your own executive-function capabilities using HI-NSA®.

Neurodiversity and change

A frequent cause of failure in change and learning processes is that we often set goals and make action plans without considering the distinct diversity of brains (more specifically, your executive or neurocognitive profile). Failure to do this can be demoralizing and makes personal and organizational change harder to realize.

Such failures can create a vicious cycle—a loss of self-efficacy (i.e., your belief in your ability to execute plans) can mean we set our sights lower. For example, in setting goals impulsively or disregarding our underdeveloped executive skills, we create a pernicious kind of self-fulfilling prophecy. Consider weight loss: "I'm going to lose 10 kilos this month by exercising four times a week (from zero) and restricting calories to 1800 (from 3100)." The likely failure happens as a result of disregarding low inhibitory-control or emotion dysregulation, and thus, people give up, wrongly believing they "self-sabotage".

> *Editorial comment*: There is an organizational parallel—overestimating change capability can lead to failed change leading to increased cynicism and lack of trust in leadership.

Change strategies based on neurocognitive development

To prevent this, we work with clients to implement a tailored mix of four strategies to develop their leadership talent and performance (see Figure III.9). By tailoring the strategies to the change (e.g., superficial change vs. deep change) and critically to individual (or team) capabilities, we create a more optimal change and learning strategy.

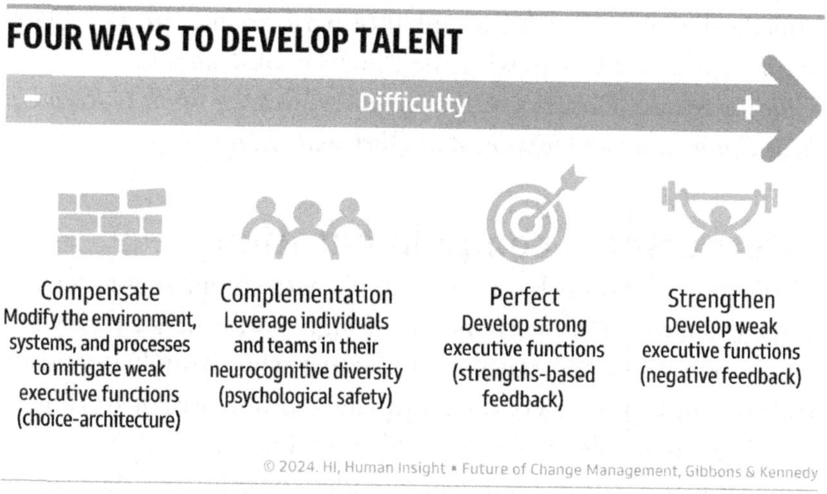

FOUR WAYS TO DEVELOP TALENT

− Difficulty +

Compensate	Complementation	Perfect	Strengthen
Modify the environment, systems, and processes to mitigate weak executive functions (choice-architecture)	Leverage individuals and teams in their neurocognitive diversity (psychological safety)	Develop strong executive functions (strengths-based feedback)	Develop weak executive functions (negative feedback)

© 2024, HI, Human Insight • Future of Change Management, Gibbons & Kennedy

Figure III.9: Four brain-based change strategies.
Development strategies should take multiple factors into account, including cognitive diversity.

Compensating weaknesses

This strategy involves making small changes in a person's environment during a change process, many of which can be automated, helping the person compensate for weaknesses with minimal effort required for their brain to consolidate new neural connections. For example, creating repetitive events with alarms in an electronic calendar to compensate for a low skill in time management and organization, or using a whiteboard and visually appealing diagrams (such as a Kanban board) to compensate for low concentration and working memory (i.e., ability to retain multiple concepts in mind). Despite its potential for immediate results and almost negligible cost or effort, this change strategy is often neglected. It is known as environmental modification in clinical neuropsychology and as choice architecture or "nudging" in behavioral economics, and its applicability extends well beyond these disciplines. Examples of pioneers in applying these ideas outside the clinical realm to enhance performance are Peg Dawson and Richard Guare with their 2016 book *Smart but Scattered. Guide to Success. How to Use Your Brain's Executive Skills to Keep Up, Stay Calm, and Get Organized at Work and at Home.*

Interpersonal complementation

This strategy involves leveraging people with complementary executive strengths. It is a cornerstone of high-performance teams and emphasizes the importance of cultivating what Amy Edmondson calls psychological safety (i.e., intimate and transparent work relationships so people feel comfortable revealing their vulnerabilities, supporting each other, and debating ideas openly). In other words, this strategy can be immediately and cost-effectively implemented in such contexts. Otherwise, it requires first focusing on repairing or building relationships and creating a work climate or culture with high psychological safety. Therefore, depending on the case, it can yield immediate or delayed and large-scale results.

Perfecting strengths

This strategy seeks excellence or mastery—moving from being good at something to being exceptional (i.e., the famous "from good to great"). It involves the brain developing new neural connections, but in a more specialized and focused manner, refining pre-existing brain circuitry and habits. This strategy can bear fruit in the medium term. Moreover, it's unclear if mastery has a limit. That's why Olympic athletes continue training, aiming to surpass record after record. This change strategy is often associated with feedback known as "positive", "appreciative," "strengths-based," or "supportive", though negative feedback can also be used for the same purpose.

Strengthening weaknesses

This strategy involves the brain creating neural circuits it doesn't have. While the perfecting strengths strategy (described above) is like growing new shoots on an existing branch, strengthening weaknesses is akin to a plant developing a new main branch from scratch. For this reason, it requires sustained effort over time that tends to yield meaningful results only in the long run. Additionally, in adults, this strategy may have a ceiling effect depending on how much a weakness can be strengthened. No matter how much we train every day for the rest of our adult lives, none of us will ever be able to play tennis like Serena Williams or football like Lionel Messi. Feedback known as "negative", "constructive," "evaluative," or "corrective," and behavioral psychology centered on habit formation focused on development opportunities are the interventions often paired with this change strategy.

These change strategies can be combined to generate early wins, medium-term achievements, and long-term accomplishments. This combination renews the motivation of individuals undergoing change or development, providing rewards along the journey that help sustain the effort required for personal development or maintaining a group or organizational change process.

One change strategy does not fit all

One-size-fits-all change strategies don't work for everyone. This is why we need to rethink how to generally address change processes in organizations. In 2023, Gibbons and Kennedy challenged the idea that change must consistently follow a specific shaped process (e.g., U-shaped with a "dip") or that the brain inherently dislikes change. People react differently to change, with some embracing it excitedly while others find it daunting. Neurodiversity plays a significant role in explaining and anticipating these diverse reactions.

Research in neuropsychology and cognitive-behavior therapy (CBT) reveals that individuals with distinct neurocognitive profiles respond to change differently. For instance, those with ADHD profiles may embrace change enthusiastically or become addicted to novelty. On the other hand, individuals with anxious or ASD profiles often view change negatively, perceiving it as a threat due to low tolerance for uncertainty. They typically exhibit initial resistance or reduced performance, sometimes following a U-shaped curve.

In sum, a person's neurocognitive profile—be it neurodiverse, neurotypical, or somewhere in between—appears to affect their response to change. Hence, adopting a universal change model or process is unwise. Instead, the future of change management involves using neurocognitive profiling to determine the most effective change approach for each individual and under specific circumstances. This idea resembles how personalized medicine sets therapeutic goals through genetic profiling.

Using existing psychometric tools and statistical models, it's possible to create an algorithm that predicts potential change trajectories for individuals. This can provide personalized recommendations for change architects leading the process and for change recipients. Leveraging this knowledge can enhance organizational change processes by tailoring them to diverse subgroups from a neurocognitive perspective.

The neurodiverse future

The neurodiversity movement envisions a future where we wholeheartedly embrace every neurological difference without exception. In other words, it seeks to establish an undeniable right for every individual, regardless of their neurological makeup, to freely express their needs and desires.

In essence, our goal is to create a neurodiversity that includes everyone.

Advancing towards greater inclusivity in neurodiversity won't be achieved by simply replacing one label with another or creating artificial divisions. True progress occurs when we go beyond broad labels and start recognizing the unique strengths and requirements of each person. This approach underscores the fact that no two individuals are alike and will facilitate the development of highly personalized strategies for more effective change management.

References

[1] Gibbons, P., & Kennedy, T. (2023). *Change myths: the professional's guide to separating sense from nonsense.* Phronesis Media: Kindle edition.

Further Reading

Armstrong, T. (2010). *The power of neurodiversity: unleashing the advantages of your differently wired brain.* Hachette Books. Kindle edition.

Dawson, P., & Guare, R. (2016). *Smart but scattered. Guide to success. How to use your brain's executive skills to keep up, stay calm, and get organized at work and at home.* Guilford Publications. Edición de Kindle.

Etchebarne, I. (2022). *Cuando la neurodiversidad excluye. 6 supuestos erróneos del movimiento de la neurodiversidad.* LinkedIn [post]. https://www.linkedin.com/pulse/cuando-la-neurodiversidad-excluye-ignacio-etchebarne/

Torralva, T., Gleichgerrcht, E., Lischinsky, A., Roca, M., & Manes, F. (2018). "Ecological" and highly demanding executive tasks detect real-life deficits in high-functioning adult ADHD patients. *Journal of Attention Disorders, 17*(1), 11-19. https://doi.org/10.1177/1087054710389988

BEHAVIORS

CHAPTER 4

Culture Change: Beyond Shared Values

by James Healy

In this essay, Deloitte Australia Principal James Healy takes aim at the predominant "creating shared values" approach to culture change. For all the importance attached to corporate values, they are hard to measure and harder to shift. Moreover, values-change does not translate reliably into behavior change—raising the possibility of misalignment and cynicism.

James shows us that behavioral science offers a more reliable and measurable effect on culture and performance, but few organizations have unleashed its power.

Culture matters—can we change it?

"Culture eats strategy for breakfast." It's the quote that launched a million LinkedIn posts.

CULTURE CHANGE: BEYOND SHARED VALUES

Widely—and erroneously—attributed to management guru Peter Drucker, its real origins in Papermaker: The Official Publication of the Paper Industry Management Association1 are slightly less prestigious. Nevertheless, the breakfast adage epitomises the consensus view that culture is critically crucial for organizational success.

That consensus is encouraged by business leaders who often ascribe their own success to the culture change they, themselves, dubiously create. Lou Gerstner famously summarised his legendary transformation of IBM in the 1980s with the words, "Culture isn't just one aspect of the game—it is the game." More recently, Satya Nadella has been explicit about the critical importance of culture in his turnaround of Microsoft and its centrality to the CEO role more generally: "I like to think that the 'c' in CEO stands for culture and it defines the success of every organisation. Our culture is at the root of every decision we make at Microsoft, and creating this culture is my chief job as CEO."[1]

Equally, when things go wrong, insiders and business media alike are quick to blame culture for corporate failure and scandal, as a cursory scan of the headlines illustrates:

- "Enron's culture fed its demise"
- "The culture of corruption at Lehman Brothers"
- "VW culture to blame for silence over emissions scandal"

These anecdotal narratives are increasingly supported by more rigorous analysis. The MIT Sloan School of Management's "Culture 500" index uses machine learning to parse more than a million anonymous employee reviews on Glassdoor.com and create a data-driven assessment of organisational culture. Putting aside for a moment the confusion of organisational culture being synonymous with "a great place to work," this research claims a direct correlation between culture and performance: The organisations recognised by employees as the best places to work delivered nearly 20% higher returns to shareholders than comparable companies over a five-year period.²

Unsurprisingly, this potential for higher returns hasn't gone unnoticed by investors. Irrational Capital, co-founded by renowned behavioural economist Dan Ariely, has taken the approach one step further. Irrational Capital calculates an organisation's Human Capital Factor (HCF) by blending publicly available information like Glassdoor reports with proprietary data sets to quantify employee perceptions of culture. Their hypothesis that a higher HCF score—and therefore a stronger culture—correlates with better financial performance has been independently corroborated by JPMorgan, who found an HCF investment strategy outperforms more traditional strategies, sometimes significantly.

> *Editorial comment*: What does a "stronger" culture mean? And, to verify findings such as this, we would have to get into the methodological weeds.

So, culture is important for organisations, but what is culture? How do organisations change it?

Adding value

The Australian Institute of Company Directors (AICD) defines organizational culture as "the shared values, assumptions and beliefs that shape the behaviour of the people involved in an organisation."

A recent Harvard Business Review (HBR) article entitled The Leader's Guide to Corporate Culture states that "culture expresses goals through values and beliefs and guides activity through shared assumptions and group norms." The same article claims that "executives are often confounded by culture because much of it is anchored in unspoken behaviours, mindsets, and social patterns."

This description is typical of the tendency to treat culture as a mysterious concept, spoken of in mystical tones, a sort of corporate lightning-in-a-bottle that everyone can see but no one can quite capture. For most 21st-century organisations, bottling that lightning simply means "shared values."

The culture-as-values approach appears simple. Firstly, define a set of abstract one or two-word, so-called values that illustrate what the organisation fundamentally stands for. This could be done autocratically from the top down; it could be crowd-sourced via co-design workshops or whole-of-organisation surveys. Once the values are agreed to, they're launched to the organisation with a flurry of communications and an avalanche of branded merchandise and collateral. Now everybody knows what the organization stands for, and culture is simply a case of ensuring everyone is "living the values." Well, that's the theory anyway. Does it work?

The shared values concept has spread like a virus, and it now permeates almost every corner of the organisational landscape. One analysis found that 397 of the Fortune 500 publish their values in their annual reports (See Figure IV.1), with companies as disparate as Tractor Supply Co., Bank of America, and Walmart proudly touting what they stand for. Outside the realm of large corporates, organisations as varied as the BBC, United Nations, and NASA publicly trumpet their commitment to living their values (which, despite the vastly different nature of those three organisations, overlap significantly)!

SAMPLING OF FORTUNE 500 ESPOUSED CORPORATE VALUES

Figure IV.1: Sampling of corporate values from Fortune 500 companies. Figure caption. Businesses talk about values non-stop and very different businesses claim similar values. But do they mean anything?

There's no doubting that shared values are a shibboleth on the organizational landscape, every executive must pay homage. Where did this approach come from?

Give me an "s"!

Organisational obsession with culture began in the late 1970s. It was driven largely by the spectre of a Japanese miracle then haunting corporate America—the fear that a resurgent Japan would soon usurp the USA as the world's economic powerhouse. Old certainties of corporate life appeared to be evaporating as the likes of Sony and Toyota rose to challenge the traditional behemoths of American manufacturing, and the world's financial centre of gravity threatened to shift from Wall Street to Tokyo. Japan's seemingly inexorable rise from the ashes of World War II to global economic superpower felt as mysterious as it was terrifying.

The Reagan administration wasn't shy to push a narrative that Japan's gains were the result of dodgy trade practices, later embarking on a full-scale trade war, infamously backed by a youthful Manhattan

construction magnate named Donald Trump. Beyond the trade polemics, there were increasing fears that something unique about Japanese culture and institutions was responsible for their success, epitomised by Ezra Vogel's book *Japan as Number One: Lessons for America*.

In response to this conundrum, McKinsey dispatched some of its finest minds, including the soon-to-be-famous Tom Peters, on a lengthy world tour to investigate the pressing topic of organizational effectiveness. McKinsey's hypothesis was that American capitalism's traditional reliance on strategy and organizational structure as the twin levers of corporate success had met its match, and a greater emphasis on "softer" elements might be required. As Peters tells the story, the work culminated in a two-day "séance" in San Francisco from which emerged the now famous McKinsey 7S model. In addition to the old stalwarts of strategy and structure were added five new pillars: Systems, skills, staff, style, and superordinate goals. The alliteration on the letter 's' was deemed crucial to the memorability of the new model, with superordinate goals the slightly desperate seventh pillar (see Figure IV.2).

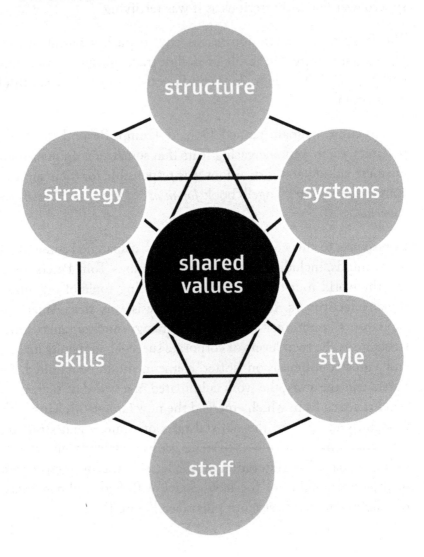

Figure IV.2: McKinsey 7S model.
More than 40 years ago, McKinsey began to tout the importance of the softer side of organization. The 7S model has stayed with us since.

CULTURE CHANGE: BEYOND SHARED VALUES

A June 1980 article in Business Horizons entitled Structure is not Organisation[4] launched the 7S model—with superordinate goals clunking away at the centre. Two of the original architects of the model, Tony Athos and Richard Pascale, included the 7S in their book published late that year, *The Art of Japanese Management*, model which claimed to provide a repeatable recipe for importing Japanese corporate culture to American companies. While this purportedly Japanese take on the future of business had some success, the 7S proved far more popular with an American accent. Tom Peters' U.S.-centric book *In Search of Excellence* was a blockbuster, selling three million copies in four years, and later voted the most influential business book of all time by Forbes.[5] It, too was built on the 7S model, but with one crucial change: Superordinate goals was replaced by shared values at the core of the model.

From these obscure beginnings as an s-shaped compromise at a McKinsey workshop 40 years ago, the idea of shared values has colonised almost every organisation on earth.

But does it work?

The dubious value of values

Today it's taken as read that shared values are the critical determinant of employee engagement and performance, and ultimately organisational success, "High-performing teams start with a culture of shared values" says Harvard Business Review. At the same time, Forbes gushes "why core values matter and how to get your team excited about them."[6]

Yet despite these pedigreed publications, evidence for these claims is surprisingly scarce. Proponents of shared values tend to cite the voluminous empirical evidence illustrating the link between culture and performance, but this relies on the somewhat circular assumption that shared values are culture. MIT Sloan researchers compared the findings from their Culture 500 analysis of employee perceptions with the values explicitly stated on company websites and horrifyingly found:

> "There is **no correlation** between the cultural values a company emphasises in its published statements and how well the company lives up to those values in the eyes of employees."[7]

Even more damningly, a 2013 study by the National Bureau of Economic Research analysed the relationship between different dimensions of corporate culture and performance across members of the S&P 500. By the usually cautious standards of academic publications, their conclusion was unequivocal: **"We find that proclaimed values appear irrelevant."**[8]

Interestingly, both sets of findings did support the thrust of Irrational Capital's Human Capital Factor approach, while proclaimed values weren't correlated with performance, employee perceptions of culture were strongly correlated with traditional performance metrics. **Culture seems to drive performance, but buzzwords on the website aren't culture.**

Do values shape ethics?

While these are academic perspectives, it's not hard to find real-world examples of shared values failing to shape culture. Integrity, communication, respect, and excellence seem like solid examples of the shared values genre: Short, simple, and clearly articulating unequivocally **good things**. Unfortunately, Enron's values weren't much use when the Texan energy trader collapsed in 2001, in the then-largest bankruptcy in corporate history. The subsequent revelations of gargantuan accounting fraud and energy market manipulation responsible for 38 rolling blackouts across the state of California didn't exactly scream **integrity, communication, respect, or excellence.**

Enron's demise had a ripple effect as global accounting powerhouse Arthur Andersen collapsed soon afterwards. Like Enron, they too proclaimed shared values of integrity, respect and excellence, and their shared values proved meaningless when it really mattered. Their belief in integrity seemed invisible as staff shredded tons of documents in a vain attempt to protect their client and themselves.

Similarly, Boeing's "Enduring Values" include integrity, quality, and safety, which won't be of great consolation to the families of 346 people killed in two separate crashes caused by the company's deliberately misleading statements to safety regulators about their new 737 AirMax planes.6 And who at Volkswagen was thinking about their website's earnest proclamation that "sustainable, collaborative, and responsible thinking underlies everything that we do," when the company installed a 'defeat device' designed to cheat on emissions tests?

Are values ever truly shared?

Of course, it's absurd to tar a whole organisation with the same brush. At the time of their respective scandals, Enron, Andersen, Boeing, and Volkswagen employed tens or hundreds of thousands of people across the globe. Only a tiny minority of employees at each company were involved in the misconduct and the vast majority were as horrified as the rest of us. Yet this precisely illustrates one of the great unspoken shortcomings of the shared values concept—any set of values just isn't that widely shared. Any organisation is composed of multiple people with different personalities, experiences, and desires and different interpretations of the wannabe shared values. The meaning of a value like "integrity" or "respect" varies from individual to individual and can vary very widely from culture to culture. This variability becomes ever more an issue in a globalised world where multinational, multicultural teams are often the norm. Try finding values that are truly shared in San Francisco, Seoul, and Saudi Arabia. The latest iteration of the Inglehart World Values Survey, which interviewed 130,000 people in 90 countries, suggests that Western values are actually diverging from other cultures' values[7] (see Figure IV.3). To counter this, shared values often tend towards trite, meaningless statements with which no one in their right mind could possibly disagree, but don't actually mean anything either.

> *Editorial comment:* We should worry that reducing culture to buzzwords and trivialities dumbs down the conversation—and may mean "values-talk" is more harmful than useful.

CULTURE CHANGE: BEYOND SHARED VALUES

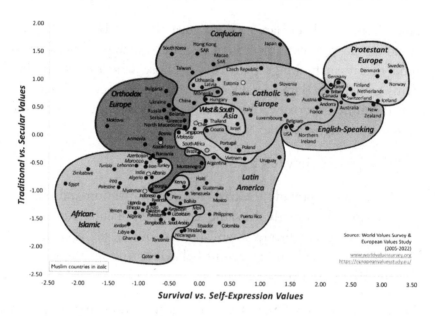

Figure IV.3: The Inglehart-Wetzel World Cultural Map.
World cultural values appear to be diverging, not converging.

While truly sharing shared values is a challenge, it's far from the only shortcoming of the values-as-culture idea. In fact, there's a much more fundamental question: **Do values *actually* influence how people behave?**

Behavioral science for beginners

According to self-declared values guru Richard Barrett, "Values stand at the very core of human decision-making." Values gurus tend to say such things (particularly those who run consulting firms called the Barrett Values Centre). Many people, even so-called culture experts, would nod along with Barrett, thinking the statement is trivially true.

Volkswagen certainly felt that way:

> "Our values provide the basis for our motivation and our decisions. It's exactly the same with Volkswagen's values.... Once they are firmly established in our heads and our hearts, they will influence our behaviour and decisions in addition to our personal values."

Yet the science of human decision making says something very different, as we're learning. Over the decade and a half since the 2008 publication of Richard Thaler and Cass Sunstein's bestselling book *Nudge*, there's been a quiet revolution as governments, big-tech companies, banks, retailers, and marketers have begun to adopt techniques and insights drawn from behavioural science to influence consumers and citizens. Behavioural science is about understanding the cognitive processes which drive human behaviour. It's a broad church encompassing a range of academic disciplines including anthropology, behavioural economics, neuroscience, psychology, and sociology. While these multiple disciplines have spawned millions of pages of academic research, three core insights are particularly relevant to organisational culture.

CULTURE CHANGE: BEYOND SHARED VALUES

Firstly, humans aren't always rational or logical, at least not in the way many of us like to believe. We're more Homer Simpson than Mr. Spock. Rather than carefully weighing up decisions with objective analysis of facts and data, real humans are often influenced by factors that cold logic would suggest ought to be irrelevant. **Wanting to do something, knowing how to do it or why to do it, isn't always enough to make us do it.** There's often a pervasive gap between intention and action, as anyone who's made a New Year's resolution can testify.

Secondly, much of our decision making is automatic. Weighing up options and deciding what to do is tiring. Our brains have evolved to conserve energy and most of the tens of thousands of decisions we make every day happen with little or no conscious thought. We use shortcuts and rules-of-thumb to make our decisions more energy efficient. Rather than solving a difficult problem like, what should I do? We'll often solve an easier problem instead like:

- What's everyone else doing?
- What have I done before in a similar situation?
- What's the easiest thing to do?"

The chances are that the question, "Does this align with my employer's values?" isn't one of those deeply ingrained rules-of-thumb. If humans too rarely analytically weigh the pros and cons of decisions, might they even more rarely weigh abstractions such as "our shared values?"

Thirdly, following from the other observations: **Environments shape behaviour.** Automatic decision making means we're profoundly influenced by context. Subtle cues in the physical environment, the behaviour of others, and the way that information is framed dramatically impact the outcome of those rules-of-thumb that make up so many of our decisions.

"Change the environment, change the behaviour" is a pillar of modern behavioural science.

These insights critically undermine the thinking behind values-as-culture. If humans aren't always logical or rational and most decisions occur at least somewhat automatically, values are much less relevant for decision making than is generally supposed. Far from Barrett's observation that "values are at the core of human decision-making," behavioural science suggests they're at best peripheral. Contrary to accepted wisdom that attitude drives behaviour, there's significant evidence that in the words of advertising legend Rory Sutherland, "behaviour comes first; attitude changes to keep up."

So, behavioural science makes clear that human behaviour is hugely complex, with dozens of factors influencing any behaviour, where explicitly stated values are among the least important of them. But what does this revelation mean for organisations?

The behavior factory

Every organisation is a combination of nested environments, unique combination of people, physical spaces, stories, symbols, processes, policies, systems, and structures make up that environment. The interplay of all those environmental factors profoundly influences the behaviour of everyone within the

organisation. In the words of Nobel laureate Daniel Kahneman, "whatever else it produces, an organisation is a factory that Manufactures judgments and decisions." As in any factory, an organisation's management has various levers they can pull to change the output from the production line.

Before they start optimising the production line, however, organisations must be clear about what they want to produce. **The first step is defining behaviours.**

This is different from defining shared values. Where a shared value is usually a vague and unspecific one-or-two word phrase, a behaviour needs to be detailed, observable, and measurable.

"Collaboration" is a typical shared value, but only by operationalizing it into a behaviour does it become useful. Who do we want to collaborate, with whom, and how? When do we want them to collaborate and to what end? Is there a specific trigger for collaboration, or a specific place—real or virtual—that we want our people to use for collaboration? And how will we know when collaboration is taking place, how will we measure it?

Once the behaviour factory managers have decided which behaviours they want to produce, and how to measure them, it's time to start pulling organisational levers to optimise the production line.

The **incentives**—both explicit and implicit—that encourage some behaviours and discourage others are a key mechanism. It's often said that "what gets measured gets managed" (Peter Drucker did say this one). What people are rewarded and recognised for is usually what they'll do. Individual key performance indicators (KPI) and metrics are crucial, particularly when they're linked to remuneration, but subtler incentives like informal recognition from leaders and peers can be just as important.

Human beings are social animals, so it's no surprise that **relationships** have a critical influence on behaviour. In an organizational context, that means the relationships formalised by the hierarchy, organisational structure, or contractual means, but it also means the informal relationships that develop between individuals. Often those informal networks are even more powerful than the boxes on the organisational chart or the bullet points in job descriptions.

Likewise, **the stories** that spread through those networks – gossip, hearsay, rumours, beliefs, accepted wisdom – can be more powerful than the formal stories told by leadership, including the much-vaunted shared values.

Tools and processes are often fatally underestimated as drivers of behaviour. Humans have a strong tendency to follow the path of least resistance; in the words of another Nobel laureate, Richard Thaler, "If you want people to do something, make it easy." To a large extent, the processes, policies, systems, and tools employed by an organisation, not to mention the physical space the organization occupies, dictate what's easy and what's hard for people to do. Even the best-intentioned staff will struggle to do the right thing in a system or a process that makes it hard to do. Even a little bit of additional friction in a process makes people less likely to complete it, a concept behavioural economist Dilip Soman vividly describes as "sludge."

These levers have a profound influence on behaviour in an organizational context. Organisations looking to change culture need to focus on defining the target behaviours they want to see, then analysing the levers they can pull to make their organisational environment more conducive to those behaviours by changing incentives, organisational structures, narratives, systems, tools, and physical environments. Merely communicating desired behaviours is no more effective than communicating shared values. Without changing the structures underpinning organisational culture, not much will change. However, it's often difficult and expensive to change these structures; pulling organisational levers and refitting the production lines in the behaviour factory takes time and costs money. What can organisations do to start shifting the dial more quickly and cheaply?

CULTURE CHANGE: BEYOND SHARED VALUES

Nudge, nudge

Given a book called *Nudge* catapulted behavioural science from the obscure fringes of academia into the mainstream, it's not surprising that much popular discussion about behavioural science focuses on so-called nudges. Arguably, the most (in)famous nudge of all involved the placement of fake flies in the urinals at Schiphol Airport, with the objective of encouraging gentleman to improve their aim. The flies were a huge success, reducing "spillage" by 80% and enabling an 8% reduction in cleaning costs! The Schiphol flies have subsequently been replicated in public toilets across the globe and are perhaps the quintessential example of the nudge in action: **A cheap, simple change to the environment that influences choice without explicitly preventing or mandating a particular behaviour.**

A PLASTIC FLY IN A URINAL IMPROVES MEN'S AIM

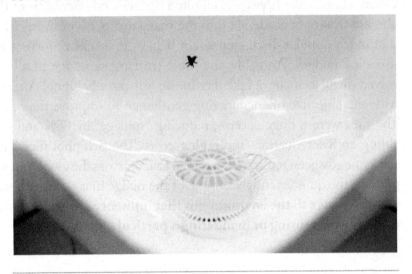

Figure IV.4: A plastic fly seems to solve the millennia-old problem of improving men's aim.
A fake fly in the urinals at Amsterdam's Schiphol Airport was enough to reduce overall cleaning costs by 8%.

Moving away from the gents' toilets, nudges have a role to play in effective organisational culture change, but it's important to understand they're not magic Jedi mind tricks. Context is key when it comes to behaviour. A nudge that works well in one context may not work at all in a different context. While there are endless applications of nudging for any given behaviour in an organisation, some real-life examples of cultural nudges at work are instructive.

SALIENT SWITCHES

Figure IV.5: Salient switches—timely, salient nudges work better.
Timely nudging for behavior change is more effective than educating or berating.

Several hazardous industries, including mining, oil and gas, and manufacturing, have made great strides in recent decades driving toward a safety-first culture. Various nudges have been successfully employed to create good safety habits. Starting every shift with a "safety share" makes safety front of mind for operators at the most important time, while those in office roles maintain a constant focus on the safety implications of their decisions by starting every meeting or conference call with their own safety shares. Stickers on bathroom mirrors proclaiming "it starts with me" provide a targeted, personalised reminder of the need to take personal responsibility for safety. In a literal twist on the environment-shapes-behaviour maxim, one construction company painted the walls a specially selected shade of "drunk tank pink" to calm their testosterone-infused workers and reduce safety (and conduct) issues, with very positive results.

Figure IV.6: Walk this way.
Figure caption. The COVID-19 pandemic introduced a whole range of nudges, many of them applying Richard Thaler's aphorism, "If you want people to do something, make it easy."

It's not just physical safety. Harvard Business School Professor Amy Edmondson's concept of psychological safety is all the rage for organisations looking to increase their people's willingness to disagree with each other and share honest, constructive feedback in pursuit of a performance culture. Successful nudges for psychological safety have included the insertion of "constructive challenge" moments such as a standing meeting agenda item, the creation of dedicated "red teams" or "black hat" roles whose job is purely to constructively challenge, and the creation of a universally understood organisation-wide catchphrase to give permission for constructive challenge. All these interventions are about making it easier for staff to have difficult conversations, by creating a specific moment, team, or role with permission to challenge, or creating a universal mechanism for anyone to ask permission to challenge. Remember Thaler's words? "If you want people to do something, make it easy".

While collaboration is one of the more popular shared values across organisations, the post-pandemic world of hybrid working is seeing increasing focus on reducing the number and frequency of meetings to allow burned-out knowledge workers to get on with their jobs. A simple plug-in for Google Meet is an effective deterrent. "Meeting Cost Calculator" automatically adds up the cost of every meeting based on the attendees and presents the would be meeting arranger with a dollar amount before they can click the send button. This intervention applies a few different nudging techniques simultaneously. Adding additional clicks to the process of sending a calendar invitation creates friction and makes it ever-so-slightly harder to set up that meeting. Showing the dollar figure puts cost front-of-mind and makes it easier to perform the cost-benefit analysis that otherwise occurs all too rarely when setting up meetings. This kind of nudge, prompting an assessment of long-term impacts at precisely the moment the behaviour will occur, is a very effective way to trigger the kind of rational, logical assessment that all too often doesn't happen in the normal course of human activity.

These are a handful of examples from a few industries, trying to solve a few specific cultural challenges, but the lessons are widely applicable. The Behavioural Insights Team coined the handy mnemonic EAST to help direct nudge design, which stands for easy, attractive, social, and timely.[9] Make the desired behaviour easier to do by removing friction, changing defaults, or simplifying language. Make the desired behaviour more attractive by using personalisation, fun messaging, funny, or unusual. Make the desired behaviour more social by encouraging social commitment and creating new social norms. Finally, use a timely reminder by prompting at the time the behaviour will occur.

The right kind of behavioral science

Blithely articulating four or five shared-value buzzwords simply isn't enough to change culture. That's just not how the human brain works. Influencing real humans that make up every organization means understanding the messy, wonderful, inconsistent, emotional, social creatures we really are, and understanding the profound extent to which environment shapes our behaviour. Every organisation is a behaviour factory where every subtle element of the environment influences millions of individual everyday behaviours which together constitute the organisation's culture. **To change the culture, change behaviour; to change behaviour, change the environment.**

Ironically enough, this was well understood by the inventors of the shared-values concept but forgotten today. As Tom Peters put it at the time:

> "In a sense, [shared values] are like the basic postulates in a mathematical system. They are the starting points on which the system is logically built, but in themselves are not logically derived. The ultimate test of their value is not their logic but the usefulness of the system that ensues."

Tragically, this critically important part of the shared values prescription has been lost. All too often, organisations misunderstand human nature and think that proclaiming their values is enough. As we've seen, it's not about the values, it's about the processes, policies, systems, and tools that make those values easy (or hard) to enact, the rituals and routines that make them front-of-mind, and the stories, rewards, and recognition moments that make clear the organisation values them.

It's all about the behaviour factory.

As Brené Brown memorably puts it, "It's better to not have values than just have bullshit up on a poster." As a prescription for organizational change, *no more bullshit posters* is a lot more useful than *culture eats strategy for breakfast*.

References

[1] *Culture eats strategy for breakfast.* (2013, May 17). Quote Investigator.

[2] *In first person: Satya Nadella.* (n.d.). People + Strategy Journal, Society of Human Relations Management (SHRM).

[3] Sull, D., Sull, C., & Chamberlain, A. (2019, June 24). *Measuring cultures in leading companies.* MIT Sloan Management Review.

[4] Waterman Jr., R.H., Peters, T.J., & Phillips, JR. (1980). Structure is not organization. *Business Horizons, June.*

[5] *The 20 most influential business books.* (2002, September 30). Forbes.

[6] Guiso, L., Sapienza, P., & Zingales, L. (2013, October). The value of corporate culture. *National Bureau of Economic Research* (NBER).

[7] The Inglehart-Welzel World Cultural Map. World Values Survey 7 (2023).

[8] Soman, D. (2020). Sludge: a very short introduction. *Behavioral Economics in Action at Rotman*, Nudgestock annual conference.

[9] Service, O., Hallsworth, M., Halpern, D., Algate, F., Gallagher, R., Nguyen, S., Ruda, S., & Sanders, M. (2014). *Four simple ways to apply EAST framework to behavioral insights.* The Behavioral Insights Team.

Further Reading

Dunbar, R., Camilleri, T., & Rockey, S. (2023). *The social brain: the psychology of successful groups.* Penguin Books UK.

Forsythe, J., Duda, J., Cantrell, S., Scoble-Williams, N., & Marcotte, M. (2024). *One size does not fit all—Deloitte human capital trends 2024.* Deloitte Consulting.

van Bavel, J., & Packer, D. (2021). *The power of us: harnessing our shared Identities for personal and collective success.* Headline.

CHAPTER 5

Evidence-Based Behavioral Science in Organizational Change

by Philip Jordanov and Beirem Ben Barrah

Sometimes, global clients or big consulting firms partner with behavioral scientists (and/or evidence-based organizational change practitioners). When this happens, old methodologies and new often collide. Moreover, internal corporate change teams may be structured differently, adding complexity that is the backdrop to collaboration.

In this chapter, Philip and Beirem walk through two case studies and illustrate how boots-on-ground behavioral scientists operate in partnership with change teams inside organizations. You'll learn about evidence-based interventions and strategies that lead the charge in change management and have been field-tested in corporate settings.

Many obstacles to overcome exist in our collective journey to make organizations more behaviorally informed. These obstacles range from improving the scientific understanding of human behavior and teaching people the fundamentals, all the way to translating insights into practical, industry-speed interventions for driving change in organizations. Effective organizational change always involves behavioral change, which means that the most effective interventions are based in behavioral science. One key element in solving this part of the puzzle is collaboration between disciplines; the other element is knowledge transfer between behavioral science teams and project teams.

In this chapter, we explore how change managers and behavioral scientists can team up—or collaborate—to make organizational change more **evidence based** and **human centric**. It's one thing to understand that all organizational change is rooted in behavioral change, but how do you truly embed the lens, tools, and approach of behavioral science into your processes? That's the focus at Neurofied, our consultancy, which specializes in applying behavioral psychology and neuroscience to organizational change management. As many behavioral practitioners have pointed out, we need to go beyond lists of biases. But how?

Paul and Tricia asked us to write this chapter after reading a draft of our book *The Dynamics of Business Behavior: An Evidence-Based Approach to Managing Organizational Change*[1]. Our goal in *The Dynamics of Business Behavior* is to bridge the gap between the disciplines of behavioral science and change management, while also providing readers with a behaviorally informed toolkit to drive positive and lasting change. It includes step-by-step guides for 18 evidence-based change interventions in areas such as stakeholder engagement, narrative and communication, and measuring change by combining scientific research. It also includes anecdotes from our consulting experience at Neurofied and insights from 40-plus interviews with industry leaders and scientists.

In early behavioral science and change management books (such as *The Science of Organizational Change*), there were few practical business examples—most came from public policy or medicine. In this chapter, we share two case studies from global corporations. We aim to show some of the forms such a partnership can take.

The organization of change

Let's assume your organization recognizes it needs to build internal change capabilities. There are many ways to go about this, and different organizational structures and cultures require different approaches. The responsible team could have names such as organizational change management (OCM), organizational development (OD), or international operations; but for simplicity, let's refer to them all as "change managers"— those who focus on facilitating the human side of change.

Every change team should ask themselves: How can our team drive change across a global organization of thousands of people (or even tens or hundreds of thousands)? From a behavioral lens, the most effective and empowering answer is to connect with others. So, a network of change leaders and agents is created. The more countries, departments, and levels of seniority these 'ambassadors of change' represent, the more likely everyone across the organization is reached.

How does behavioral science strengthen basic organizational change ideas such as a change network? Even though this approach is valid and time-tested, the intention-action gap poses a key challenge to its effectiveness. That gap between individuals' intentions and their actual behaviors undermines the potential of these networks. Stakeholders across the organization might be committed to dedicating, say, 10% of their capacity to driving change, but unless the necessary behaviors are prioritized, clarified, and reinforced—and barriers adequately addressed—there's a strong chance that intentions won't translate into any meaningful action. (Later in this chapter, we introduce an evidence-based intervention to help mitigate this risk.)

Let's make another assumption: The lead of the global change management team decides their change managers and agents can benefit from a deeper understanding of human behavior. How do they go about doing so? There are many insights and interventions rooted in behavioral science that help during organizational change. Training change managers in behavioral principles is a good first step to raising awareness, but how do you go beyond awareness and understanding to practical mastery?

We have found that organizations take different approaches to leveraging our behavioral expertise in order to build internal change capabilities. Each approach has its own strengths, weaknesses, and pitfalls. Let's focus on two primary models for collaboration between change managers and behavioral scientists: Centralized and functional structures.

A tale of two change styles

In exploring the collaborative efforts between change managers and behavioral scientists, we focus on two clients, both global healthcare titans—Company A and Company B. Our role as a behavioral business partner is to cultivate the mindset and capabilities for driving meaningful and sustained change, leveraging insights and interventions from behavioral science. While the purpose and ambitions of these partnerships are similar for both companies, the approach varies given the differences in how change is structured at each company.

Centralized structure

Company A, with a history spanning almost a century and a half, shows a clear awareness of the necessity for continuous renewal and adaptation. They place organizational change at the center of their strategy. The global lead of change and communication oversees a team of about 30 regional change leaders spread across the globe. Regional change leaders collaborate closely with local change managers to facilitate change initiatives.

Surrounding this core team is a broader network of over 300 change agents—employees dedicating a portion of their time to change efforts, who at Company A are known as the Leading Change Community. While this centralized structure is primarily designed to embed strategic changes down to frontline teams, it also incorporates bi-directional feedback mechanisms, which allow for grassroots change initiatives to emerge and be recognized through the community.

Functional structure

Company B presents a contrasting approach to managing its organizational changes. Transitioning from a family-owned enterprise to a high-growth global healthcare leader, they have experienced a big shift in team dynamics and operating models, moving from a regional to a global focus. Only recently did this transition call for the appointment of a dedicated organizational change lead to oversee international operations. Now, with OCM leads in each of their seven regions, Company B is working towards two key objectives: Refining its change creation and implementation processes and elevating organizational change management beyond communication and training. This transition aims to place change management, readiness, and leadership at the core of their strategic agenda.

To have a tangible impact on behavior, behavioral scientists must adapt their interventions and experiments to fit the context in which organizational change is structured.

Role of behavioral scientists

Generally, behavioral scientists apply principles from psychology and other behavioral science fields to influence and understand human behavior in various contexts. In organizational change, this entails designing and implementing interventions to drive behavior changes that are aligned with an organization's unique goals, structure, and culture.

How do these two different change structures (from Companies A and B) affect the role of behavioral scientists? In a **centralized** setting like Company A, behavioral scientists often get to engage in strategic planning alongside the organization's change leaders. Their focus is on developing interventions that cohere across regions and departments. This helps ensure that change initiatives are consistent and aligned with overarching strategic goals, but it requires a nuanced understanding to address local needs and contextual differences within the organization.

On the other hand, in a **functional** structure like in Company B, the role of behavioral scientists tends to be more dynamic and adaptable. They work closely with different departments and teams, such as HR business partners and regional leaders, tailoring interventions to meet the specific challenges and shifting priorities of each unit. This second, tailored approach allows for effective management of rapid changes and growth within the organization but can make it more challenging to ensure consistency across initiatives.

Companies A and B also differ in their level of maturity, meaning different challenges.

Company A's centralized structure and maturity, with an established market presence and product lines, called for a coordinated, more consistent approach to change. However, this can instigate a **'red-tape crisis'** or challenge of bureaucracy.

Comparatively, Company B's functional, decentralized structure was aligned with the needs of a growth-stage company, which demands adaptability to evolving market conditions and internal shifts. However, this can lead to a **'control crisis'** or challenge of maintaining consistency and coherence across all initiatives.

In the centralized structure of organizations like Company A, behavioral scientists run into the difficulty of ensuring sufficient engagement across all themes and departments. Despite strategic alignment at the top level, translating these strategies into sustainable behavioral changes at the work-floor level remains a tough nut to crack. As a client once beautifully captured: "This gap between strategic intent and practical implementation often means that well-conceived initiatives fail to achieve lasting impact." Behavioral scientists must juggle these complexities and recognize that adoption is a core responsibility in managing change, not just an end-user issue.

Working within Company B's functional structure, behavioral scientists will find that achieving local impact and engagement is more straightforward, thanks to closer alignment with specific departmental needs. However, sustaining sponsorship and building behavioral change capabilities at every organizational level is much more difficult, as it requires you to connect with more stakeholders and teams. One further challenge in such decentralized settings is facilitating cross-departmental learning and enabling change at scale. Here, behavioral scientists ensure that interventions don't just address immediate departmental needs but also contribute to broader organizational learning and development.

Evidence-based change interventions

How do behavioral scientists achieve the above in practice? We turn to techniques called evidence-based change interventions (EBCIs). These interventions are grounded in scientific research and have been validated through real-world testing. EBCIs are designed to change behavior within complex organizational contexts and are essential tools in the toolkit of every change manager. Most EBCIs we design are rooted in behavioral science, but the most effective augment those roots with insights from other disciplines such as systems thinking and/or neuroscience.

In our book *The Dynamics of Business Behavior*, we detail a step-by-step approach for eighteen different EBCIs across six change areas often used in our client projects:

1. Planning and risk management,
2. Narrative and communication,
3. Leadership support,
4. Stakeholder engagement,
5. Measuring change, and
6. Learning and development.

Collectively, these six areas enhance decision making, foster increased engagement and a culture of psychological safety, promote continuous learning, and improve returns on change initiatives (see Figure V.1).

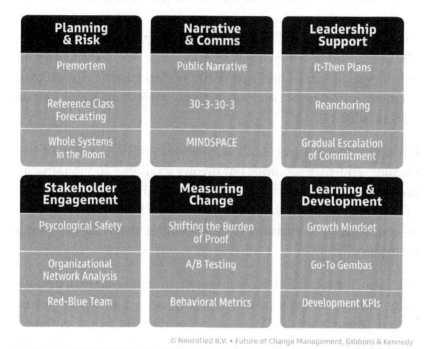

Figure V.1: A taxonomy of evidence-based change interventions in six categories.
Evidence-based change interventions are useful across all aspects of change management.

Oftentimes these EBCIs run counter to senior leader habitual practices. For example, most change models (and most change leaders) assume that large-scale change initiatives should be top-down. Whereas two EBCIs in our book—organizational network analysis (ONA) and whole-system-in-a-room (WSR)—shift the focus toward emergent and bottom-up change. ONA maps the informal hierarchy and social networks within the organization, and WSR is about bringing various stakeholders together to drive the change from within instead of by command and control.

This focus shift suggests two different kinds of augmentations to a change expert's toolkit. One, interventions based on the emergent change paradigm, and two, interventions grounded in evidence. The latter, evidence-based interventions also demand the debunking of pseudoscience and myths rife within current change management frameworks (such as in *Change Myths* published by our editors in 2023).

Change management is both science and art. EBCIs are not simple plug-and-play solutions but involve a tailored and iterative approach that leaves room for 'artistic' customization. In the context of Companies A and B, the application of EBCIs differed because of the different structures through which they execute change management.

We have learned a lot pioneering these newer approaches. Initially, our behavioral science work emphasized an in-depth analysis of cognitive biases. We now focus on EBCIs, the core tools of our craft. And in most cases, we're spending more time and words on practical change topics than on explaining the behavioral science behind it. Behavioral insights and interventions can be game-changing, but scientists should take special care to empathize with the needs of change managers and leaders with which they partner.

Sometimes, teams need to understand their biases. For example, when they observe destructive team dynamics, it can be very useful to know which biases might be at play. But other times, biases are just a starting point. The goal is usually positive and lasting change and we have found that EBCIs are tools just as useful for change managers and leaders as they are for behavioral scientists.

Collaboration in practice

To illustrate the practical application of EBCIs in different organizational settings, we focus on two specific techniques (covered in more depth in our book): The premortem and 'if-then' plans. We chose these two interventions because they are relatively simple to apply and have many positive spillover effects in terms of stakeholder engagement and leadership development.

Premortem

Premortem interventions, conceived by psychologist Gary Klein—a leading expert on intuitive decision making[2]—are designed to anticipate potential risks and develop proactive strategies for mitigating them. Those interventions create a structured scenario where team members envision a future where a project has failed terribly and then work backward to identify which things led to such an outcome. While usual risk analysis relies on foresight to predict and mitigate potential risks, the premortem technique leverages hindsight in a prospective manner. In short, it's a formal way of putting to good use the refrain, "We should have seen this coming!"

If-then plans

If-then plans were conceived by psychologist Peter Gollwitzer[3] and are aimed at shaping and reinforcing new behaviors, turning them into scalable habits across an organization. This intervention involves creating specific action plans—if-then statements—that link situational cues to desired behaviors. For example, "If I encounter [specific situation], then I will perform [specific behavior]". This approach is particularly effective in helping individuals and teams adopt new behaviors in a consistent manner, making it a valuable tool for fostering behavioral change in various organizational contexts. It's one of the EBCIs that is easy to experiment with but hard to master. Remember the intention-action gap we discussed earlier? If-then plans are directly aimed at mitigating this risk and are therefore useful for change champion networks.

> *Editorial comment*: These are also sometimes called implementation triggers in the literature on habits.

For both of these interventions, the distinction between centralized and functional organizational structures significantly influences their implementation.

Premortems in action

In centralized organizations like Company A, the premortem intervention typically uses a train-the-trainer approach. We equipped central change leaders with the necessary skills to conduct premortems across different regions. This strategy brought a diverse range of insights to the forefront, leading to a more comprehensive evaluation of potential risks and blind spots. The key for behavioral scientists in this setting is to guide these leaders in effectively utilizing the premortem technique, emphasizing the importance of varied perspectives within the group. We found that diversity within the group generally improves the quality of outcomes, but in centralized structures, this is even more important as it captures more perspectives, bringing in diverse contexts and nuances of risk across the organization.

In functional organizations such as Company B, the premortem process tends to be more hands-on and directly engaged with departmental teams. Here, facilitation centers on addressing specific, localized challenges. The behavioral scientist plays a crucial role in directly guiding these teams through a premortem process, ensuring that the intervention is tailored to address their immediate and specific needs. This type of involvement also means the behavioral scientist will have direct access to insights and can more proactively look for opportunities to cross-pollinate learnings among stakeholder groups.

If-then plans in action

Shifting to if-then plans, their implementation in centralized organizations like Company A involves translating strategic objectives into scalable habits that resonate across different regions and departments. In one organization we worked with, the CEO outlined four strategic themes focusing on leadership, empowerment, decision making, and time management. We collaborated closely with regional leaders to develop if-then statements that embodied these themes. For instance, regarding decision making, an if-then plan can be formulated as: "If I face a decision that impacts my team, then I will consult with at least two team members before finalizing it."

In functional structures, if-then plans need to be agile and highly customized to each department's immediate challenges. The intervention focuses on creating actionable, context-specific plans for fostering new behaviors. In our experience, behavioral scientists in this context concentrate on working with local managers on crafting if-then plans with very specific "if" conditions, unique to local challenges. Local managers are then coached to communicate and role model these if-then plans in an authentic and consistent way and guide their teams to adopt them in day-to-day processes and ways of working.

In short, EBCIs like premortems and if-then plans demonstrate two practical, evidence-based ways to enhance change management. Premortems help teams think ahead and mitigate future risk, while if-then plans help stakeholders modify habits step by step. Both can be adapted to the way in which an organization approaches change, whether centralized, functional, or somewhere in between. This is where change managers and behavioral scientists can really make an impact, collaborating to tailor these interventions to different contexts.

Conclusion

Whether you are a behavioral scientist, a corporate change manager, or leader, we hope you have found this chapter insightful. There is no way to present a comprehensive overview of collaboration of this type in a single essay, but we did touch on many of the key components and our lessons learned. Some of the questions we attempt to answer are:

- How does behavioral science add value to organizational change management?
- What is the role of behavioral scientists in this process?
- How is change structured at different organizations?
- How do these structures affect the role of behavioral scientists?
- What are EBCIs and how does their implementation differ across organizational structures and cultures?

For scientists, we hope we provided a clearer idea of how to leverage expertise and develop skills in a way that provides practical value instead of abstract knowledge. The important thing to remember is the behavioral scientist's goal is to guide and support the change process through a human-centric lens with evidence-based tools. Change managers won't often have had the time to study human behavior as deeply as a behavioral scientist, but they have contextual knowledge and intuition.

For change professionals and leaders, it's the other way around. You are at the core of actual organizational change, and no number of scientific insights and interventions will fully replace the intuition built up from years of experience. However, scientific insights can help you improve and refine your efforts. Collaborating with scientists empowers you to drive change without having to worry too much about the fundamentals of how humans think, decide, and act.

Together, these two groups create an effective team whose combined value is more than the sum of its parts. Collaborating might not always be easy, as some intuitive and long-held beliefs of change managers are overthrown by scientific insights (e.g., find plenty of great examples in *Change Myths* our editors). Likewise, scientists might find it tough to let go of some validity and robustness in exchange for speed, operational or cultural fit, and other forms of practical business value. But if we can push through these barriers together, the future is bright.

We hope this chapter inspired you to analyze and improve your organizational change management style so that you might explore, experiment with, or extend this type of collaboration and make your change efforts more behaviorally informed.

To learn more about EBCIs, have a look at our book *The Dynamics of Business Behavior*. We also highly recommend Paul and Tricia's previous works, as they have played an important role in getting us to where we are.

Before reading the next chapter, we challenge you to set at least one experiment in motion!

References

[1] Ben Barrah, B., Jordanov, P. (2024). *The dynamics of business behavior: an evidence-based approach to managing organizational change* (1st ed.). Wiley.

[2] Gollwitzer, P.M. (1999). Implementation intentions: strong effects of simple plans. *American Psychologist*, 7(54), 493.

[3] Klein, G. (2007). Performing a project postmortem. *Harvard Business Review*, 9(85), 18-19.

Further Reading

Gibbons, P., & Kennedy, T. (2023). *Change myths: the professional's guide to separating sense from nonsense*. Phronesis Media: Kindle edition.

Gibbons, P. (2019). *The science of organizational change: how leaders set strategy, change behavior,a nd create an agile culture* (2nd ed.). Phronesis Media.

Sibony, O. (2020). *You're about to make a terrible mistake! How biases distort decision making and what you can do to fight them*. Swift Press.

Soman, D., & Yeung, C. (eds.) (2020). *The behaviorally informed organization*. University of Toronto Press.

Weisbord, M., & Janoff, S. (2010). *Future search: getting the whole system in the room for vision, commitment, and action*. Berrett-Koehler Publishers.

CHAPTER 6

Applying a Behavioral Science Lens to Human Resources

by Scott Young

At the core of every human resource (HR) policy is the need to influence employee behavior, perhaps such as cybersecurity or diversity, equity, and inclusion (DEI). Moreover, HR's essentials, which include recruitment, retention, rewards, and learning, have behavioral components.

When behaviors do not align with policy intentions, or worse revolt against them, the costs in financial, risk or morale may be significant.

A reason behaviors don't change is that they rely on one-dimensional reasoning and a limited toolkit rooted in 20th-century psychology. They assume that when employees fail to adapt to a new policy, process, or strategy, they:

- *Do not know enough about it (and need more information);*

- *Feel threatened by it (and need coercion or more information to "fix");*
- *Disagree with it (and information will convince them); or*
- *Are not incentivized correctly.*

These assumptions ignore what behavioral science has taught us about human behavior and decision making during the last 20-or-so years. Today, behavioral science tools are beginning to vastly improve HR outcomes. In this chapter, Scott Young, a leader in the behavioral science field, shares his expertise in the HR domain.

What is behavioral science?

Behavioral science is the study of human behavior and decision making. It is a newer field that combines rigorous academic research—in behavioral economics, psychology, neuroscience, sociology, and more—with practical application. The field was popularized by books such as *Nudge* (2008) by Nobel prize winners Richard Thaler and Cass Sunstein and *Thinking, Fast and Slow* (2001) by Daniel Kahneman.

A foundational concept of Kahneman's that helps explain why behavior change efforts fail is **system-one and system-two thinking.** Traditional economics erroneously taught us that humans are rational beings who consciously sort through all available information, weigh options, and choose the path that maximizes our satisfaction and utility. However, this is rarely the case. We often rely on largely subconscious, fast, automatic system-one thinking to navigate through life. We draw on mental shortcuts—called heuristics—to shorten decision-making time and expend as little cognitive energy as possible. In short, we naturally gravitate towards the easiest course of action, which is typically our established habits and routines. These habits of mind and behavior are an oft-neglected reason why change proves difficult.

Traditional HR tools are often designed assuming system two is the primary way of thinking, or at least will prevail, where employees will devote mental energy—think carefully, methodically, and rationally—and behave accordingly. For example, incentives such as end-of-year bonuses assume that employees will keep this long-term benefit in mind and consistently act in ways that will move them closer to it. This ignores the reality that humans tend to focus on the present and shorter term rather than the longer-term future. We fall victim to **present bias** and **hyperbolic discounting**, choosing quicker satisfaction over intangible, longer-term rewards. This is one reason why it is so difficult for many of us to exercise regularly or eat healthily—because the 'sacrifice' is immediate, while the potential reward often feels remote and distant.

Kahneman's concepts of fast and slow thinking hold profound implications for influencing employee behavior during organizational changes. It exposes why simply giving employees information through a training session or presentation rarely leads to behavior change but rather overwhelms them with concepts.

Be specific about desired behaviors

The first step to lasting behavior change is to descend from the 30,000-foot view of behavior and down to the ground floor of individual actions. Though simple, several critical questions (not unique or radical) are too rarely asked (see Figure VI.1).

CRITICAL BEHAVIOR CHANGE QUESTIONS FOR ANALYSIS

Whose	What	When	How
behavior do we expect to change?	exactly is person or group expected to do?	is the behavior expected to change?	is the behavior change measured?

© 2024, Future of Change Management, Gibbons & Kennedy

Figure VI.1: Critical questions to answer during behavioral change analyses. Many HR initiatives fail because the essential behaviors are not clearly specified.

These questions may seem ordinary, but clients implementing "sustainability" or "diversity" initiatives often fail to make those terms behaviorally specific, and hence actionable. Moreover, by tightly defining behaviors this way, we can design an evidence-based intervention—one grounded in science, not supposition or hope.

Behavioral intervention using EAST

"If we want people to do something, make it easy."

RICHARD THALER, NOBEL LAUREATE

A more general, evidence-based framework called EAST (ie., easy, attractive, social, and timely) is a useful tool (see Figure VI.2). Though simple, this framework helps HR professionals design efficient, effective interventions—sometimes one where only a small input makes a big difference.

EAST FRAMEWORK

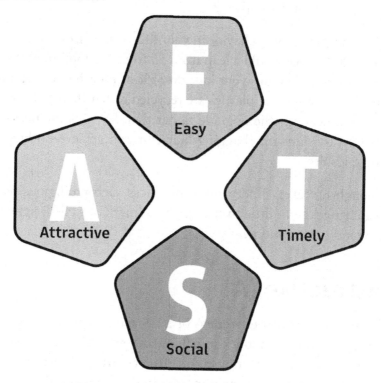

Figure VI.2: EAST stands for easy, attractive, social, and timely.
Using EAST, behavioral interventions can be designed efficiently for maximum effect.

Easy (E)

Interventions to make a behavior **easier** can involve reducing the number of steps it takes to perform or simplifying instructions. An engineering consultancy, for example, was able to increase the percentage of staff that completed demographic disclosure forms from 55 to 85% by simplifying the process and removing barriers to survey completion, including time constraints and employee concerns about data privacy.

Consider, a specific behavior in your life, business, or society. How much difference would a leap from 55 to 85% adoption of this behavior make? Perhaps your kids do 85% of their homework rather than 55%, or 85% of plastics are recycled rather than 55%, or 85% of customers buy the upgrade rather than 55%. Some businesses spend vast sums on changing behavior and achieve nothing like these results.

Though not every intervention produces such prodigious results, big changes from little tweaks are a hallmark of well-designed behavioral interventions.

Attractive (A)

Enhancing the **attractiveness** of a behavior might entail personalization, which is more compelling and attractive to people than more general interventions. For instance, truck driver safety and fuel efficiency are important metrics in the transportation industry. By focusing on building personalized feedback, messages, and recommendations into driver performance software, improvements in these metrics were significant.

Employee physical activity is important for fitness and stress reduction. One company gamified fitness with free FitBits, peer and group competition, and customized, personalized feedback. Groups who received personalized feedback significantly outperformed the others.

Social (S)

Leveraging the **social** components of change is another powerful and often underleveraged driver.

Communications efforts and incentives are often focused solely on the individual, while neglecting the potential power of colleagues acting together and network effects of group norms. Understanding (and leveraging) how behaviors spread through social networks within organizations is crucial, as is incorporating behaviorally-informed messaging to reinforce, for example, a culture of psychological safety.

A multinational manufacturing corporation found that encouraging departments to take up positive behaviors together—comparing their performance to that of other departments—increased the uptake of employee physical and financial wellness programs.

Timely (T)

Finally, it's important to **time** efforts properly. On one level, this involves prompting people when they are most likely to be receptive to change. On another level, it involves the timing of communications—and the importance of complementing in-depth training (which is often removed from the moment of action) with follow-up "in the moment" reminders, ideally built into important processes.

For example, so-called fresh start events such as birthdays, the beginning of the year, or the first day of a new job are examples of dates employers could leverage to reach their audience in a timely manner. These events mark new beginnings during which people are more open to setting goals and changing habits and routines.

New tools for people management

While behavioral science can produce specific changes in employee behavior through targeted projects and interventions, it can also catalyze more foundational changes when integrated into an organization's core processes, many of which are owned by HR and make up the "wiring" of an organization.

Behavioral science offers tools to help with these core processes, including recruitment, retention, compliance, and performance management.

DEI recruitment

While many companies and individuals have positive intentions related to increasing workforce diversity, it often fails to translate to recruitment behaviors, partly due to unconscious biases. Hiring managers may naturally and unconsciously gravitate toward applicants whom with they are most comfortable and share characteristics, potentially overlooking other equally qualified candidates. Unconscious biases also affect judgments made during performance management.

Behaviorally-informed recruitment processes can help debias recruitment systems (or at least minimize any bias). Some practices are well established, such as removing names and other personal identifying information from resumes, on which behavioral science coupled with technology can improve. An app called Applied has been designed using behavioral science and data science to reduce bias, for example, to improve the quality of hires and increase their diversity. One way it does this is by randomizing candidates' answers and requiring hiring managers to review all answers to each application question simultaneously, as opposed to the more traditional approach of reviewing each person's application one at a time. This helps reviewers focus on qualifications rather than identities, preventing bias toward some candidates over others.

Retention

Burnout and high turnover are challenges in many organizations, particularly in fast-paced and stressful professions, such as nursing and financial services. Typically, companies try to encourage employees to stay through tangible, rational appeals—perhaps offering more money, more vacation time, or other bribe-like incentives. While these efforts are important and well-intended, they often come too late and/or have only limited effects. By the time someone is burnt out, offering rewards to get people to stay usually only works in the short term: The deeper roots of burnout and turnover remain unaddressed.

One behavioral science tool allows appealing to the social components of a job to increase retention. Interviews with 911 dispatchers, for example, revealed they felt isolated and undervalued compared with other emergency service workers. Promoting belonging and connection between dispatchers by creating bonds and social norms lowered self-reports of burnout. Resignations decreased by more than half.

Compliance

Despite investments to promote and track employee compliance, adherence to company policies is often limited. Most organizations assume that this is the result of employees not fully **knowing** the rules and expectations, or perhaps not **caring** enough about them. This is the classic **information deficit hypothesis**: If we just give people more information and convince them of a policy's importance, then their behavior will change.

This thinking has led to compliance programs that emphasize education and persuasion. Yet while training is often necessary, providing people with information does not consistently lead to behavior change. We've seen this repeatedly on issues ranging from diversity and inclusion to cybersecurity.

Bluntly, walking people through the processes and expectations and telling them that something is important generally fails to make a dent in actual practices and outcomes. This is often because the majority of training is temporally removed from the actual activity. If an employee is trained on a process but does not have to undertake it for weeks or months, the chances that they will remember and act appropriately are slim.

The majority of non-compliance is inadvertent. Most employees want to do the right thing. They do not actively disagree with a company's policies but may not follow them because they are too complex, confusing, or difficult to follow—or go against their established comfortable habits. In these cases, **simplification** is a more likely solution than further education or persuasion.

Similarly, there is evidence that most people who commit fraud and other forms of malfeasance are not bad actors who set out to do so from the start. Instead, it often begins with a single unethical act—perhaps to cover up a mistake—and then snowballs over time as the person falls deeper behind expectations. This means there is an opportunity to develop behavioral interventions to help reduce the likelihood that relatively small problems don't lead to larger acts of deception and malfeasance.

Essential tips for application

To be effective leaders must find the right opportunities to maximize the impact and value of behavioral science, position the behavioral lens properly, and adopt an experimental mindset.

Focus on intention-action gap

Behavioral science interventions are most effective in situations in which people are already open to change but have barriers standing in the way (such as established habits). Fortunately, positive employee intent exists in many HR areas. People generally want to adapt and continue performing well, and they want to take advantage of new opportunities.

Organizations should target behavioral science projects around these situations, rather than in trying to drive change where underlying motivation is limited.

Behavioral science is not a panacea. It may not be effective if going against the grain of employees' intent or existing incentives. If we are incentivizing one form of behavior and simultaneously nudging people in a different direction, we are not likely to be successful.

Complement traditional tools

There will always be a need for traditional tools, such as education and financial incentives, in order to give people the ability and the intent to change behavior. Behavioral science should not be positioned as a replacement for these efforts. Instead, it is leveraged to enhance and complement them by taking human nature into account. In other words, behavioral science should be viewed as an extra dimension or a new lens that can make existing investments—in training, recruitment, etc.—more effective in driving intended actions and outcomes.

Experiment and test

Behavioral science is especially valuable in its measurability—they are doing it, or they aren't. In contrast, thoughts and feelings can only be measured indirectly and unreliably.

To build measurement into a process, use TESTS (see Figure VI.3).

- **Target**: Choose the behavior to focus on and how to measure it.
- **Explore**: Research the context in which the behavior takes place.
- **Solution**: Generate ideas and refine them based on feasibility and potential impact.
- **Trial**: Test the solution and evaluate results.
- **Scale**: Depending on the results, scale the solution more widely.

TESTS FRAMEWORK

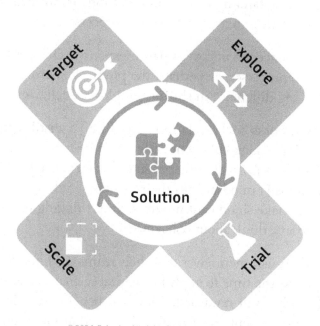

Figure VI.3: TESTS stands for target, explore, scale, trial, and solution. TESTS is an efficient checklist for building measurement into behavioral science interventions.

Evaluating the results of behavioral interventions is always valuable, as it ensures that organizations learn from experience and can scale successful interventions for maximum effect.

Behavioral approach to HR

Applying a behavioral lens to HR and organizational change can be powerful. It helps drive specific changes in employee behavior and augments traditional approaches to retention, compliance, and other HR challenges. This lens can ensure HR policies are guided by a deep understanding of human behaviors; it can help people successfully adapt to new structures and processes.

Behavioral science can uncover small, inexpensive changes to communications, processes, or perhaps the physical environment that can make a big difference in shifting employees' behavior.

While frameworks such as EAST and TESTS distill core concepts, applying behavioral science effectively isn't as simple as adding a few nudges to an existing program. Leaders need to systematically build this mindset into their thinking and organizational processes to make significant progress. Stop thinking first or only about mindsets, think about behaviors!

Leaders who push themselves and their teams to think differently—and invest the time to apply behavioral science consistently—are likely to be well rewarded. They will benefit from increased employee engagement and retention, better adoption of organizational changes, and ultimately, more effective HR teams.

Further Reading

van den Aker, M. (2023, November 19). *Interview with Scott Young.* Money on the Mind. https://www.moneyonthemind.org/post/interview-with-scott-young

Young, S. (2023 April 25). *How (& why) to start infusing your company with behavioral science.* Ethical Systems. https://www.ethicalsystems.org/how-why-to-start-infusing-your-company-with-behavioral-science/

Young, S. (2020, June 22). *Finding opportunities to apply behavioral science for good in the private sector.* Behavioral Scientist. https://behavioralscientist.org/finding-opportunities-to-apply-behavioral-science-for-good-in-the-private-sector/

TOOLS

CHAPTER 7

Behavioral Science Tools for the Change Professional

by Robert Meza and Paul Gibbons

We aren't good at changing behaviors: Not personal behaviors, such as wellness habits and routines, nor work behaviors, such as inclusion and good cybersecurity practices. We aren't good at aligning workplace behaviors with cultural values, such as innovation or psychological safety, nor are we effective in society, changing behaviors that benefit us all such as vaccination, pro-climate, obesity, and addiction behaviors.

Behavioral science tells us how we can do better in theory, but until now change practitioners have been left to fend for themselves—to come up with their own practical applications and tools based on whatever knowledge they have of behavioral science. Robert Meza, founder of Aim for Behavior (aimforbehavior.com), offers an evidence-based, online toolkit based on his work with private sector organizations and governments. In this chapter, we show how the complexity of human behavior change is enabled by a few of these tools.

The Scylla of complexity, or the Charybdis of over-simplification?

In the *Odyssey*, Homer's Odysseus risks death, navigating between two sea monsters, Scylla and Charybdis. Our monsters are over-complexity and harmful oversimplification.

Change, from **personal** change through **corporate** change to **societal** change, demands behavior change. This seems so obvious that it does not bear saying, yet in most instances, particularly in the organizational change arena, we are over-reliant upon a model Paul sometimes calls "persuade then pray." We pray that our superb communication, education, influencing, and inspiration will **persuade**, and **pray** that behaviors will inevitably follow. But, as the Chinese proverb says, "Talk does not cook rice."

Persuade-then-pray is how, for example, most diversity training is expected to work. However, according to Harvard sociologist Frank Dobbin, "Hundreds of studies dating back to the 1930s suggest that antibias training does not reduce bias, alter behavior, or change the workplace."

A more recent example of persuade-then-pray, from 2023, is the last step in the culture change methodology of a $10 billion consulting firm: Communicate New Values. If that is the final step, or where their culture change approach ends, they had better be very effective at praying for new behaviors.

That is also why your authors roll their eyes every time another n-step, recipe-book change model, divorced from research in behavioral science (such as ADKAR), claims to be the answer to changing behavior. Those simplifications require a lot of prayer.

Those old approaches represent the Charybdis, grotesque oversimplifications that produce limited results.

As simple as possible, but no simpler

Human behavior is a complex thing. Who knew?

Recall the paraphrase of Einstein that everything should be as simple as possible, but no simpler. How is this balance accomplished when it comes to behavior change?

The good news is that in just the last decades, many researchers (including from University College London) have produced a taxonomy of behavior change techniques (BCTs) from behavioral science research across a panoply of disciplines (including, for example, medicine, public health, social policy, and environment).

The bad news, the Scylla to extend our metaphor, is that there are 93 of them (see Figure VII.1), which although comprehensive can also be difficult to navigate.

Taxonomy of 93 Behavior Change Techniques (BCT) in 16 Clusters

(alphabetical order by cluster and technique)

1. Antecedents	6. Goals and planning	11. Repetition and substitution
1.1 Add objects to environment 1.2 Avoid and/or reduce exposure to cues 1.3 Body changes 1.4 Distraction 1.5 Restructure physical environment 1.6 Restructure social environment	6.1 Action planning 6.2 Behavioral contract 6.3 Commitment 6.4 Discrepancy between behavior and goal 6.5 Goal-setting behavior 6.6 Goal-setting outcome 6.7 Problem solving 6.8 Review behavior goal(s) 6.9 Review outcome goal(s)	11.1 Behavior substitution 11.2 Graded tasks 11.3 Habit formation 11.4 Habit reversal 11.5 Overcorrection 11.6 Practice and rehearsal 11.7 Target behavior over-generalisation
2. Associations	**7. Identity**	**12. Reward and threat**
2.1 Associative learning 2.2 Cue signaling reward 2.3 Exposure 2.4 Prompts and cues 2.5 Reduce prompts and cues 2.6 Remove aversive stimulus 2.7 Remove reward access 2.8 Satiation	7.1 Framing and reframing 7.2 Identification of self as role model 7.3 Identity associated with changed behavior 7.4 Incompatible beliefs 7.5 Valued self-identity	12.1 Future punishment 12.2 Material incentive 12.3 Material reward 12.4 Non-specific incentive 12.5 Non-specific reward 12.6 Outcome incentive 12.7 Outcome reward 12.8 Self-incentive 12.9 Self-reward 12.10 Social incentive 12.11 Social reward
3. Behavior comparisons	**8. Natural consequences**	**13. Scheduled consequences**
3.1 Behavior demonstration 3.2 Information about approval from others 3.3 Social comparison	8.1 Anticipated regret 8.2 Consequence salience 8.3 Emotional consequence monitoring 8.4 Information about emotional consequences 8.5 Information about health outcomes 8.6 Information about social and environmental consequences	13.1 Behavior cost 13.2 Punishment 13.3 Reduce reward frequency 13.4 Remove punishment 13.5 Remove reward 13.6 Reward alternative behavior 13.7 Reward approximation 13.8 Reward completion 13.9 Reward incompatible behavior 13.10 Situation-specific reward
4. Covert learning		
4.1 Imaginary punishment 4.2 Imaginary reward 4.3 Vicarious consequences		
5. Feedback and monitoring	**9. Outcome comparison**	**14. Self-belief**
5.1 Behavior feedback 5.2 Behavior monitoring by others without feedback 5.3 Behavior self-monitoring 5.4 Behavioral outcome feedback 5.5 Behavioral outcome monitoring by others without feedback 5.6 Behavioral outcome self-monitoring 5.7 Biofeedback	9.1 Comparing imagined future outcomes 9.2 Credible source 9.3 Pros and cons	14.1 Focus on past success 14.2 Mental rehearsal of successful performance 14.3 Self-talk 14.4 Verbal persuasion about capability
	10. Regulation	**15. Shaping knowledge**
	10.1 Conserve mental resources 10.2 Paradoxical instructions 10.3 Pharmacological support 10.4 Reduce negative emotions	15.1 Behavioral experiments 15.2 Behavior-performance instruction 15.3 Information about antecedents 15.4 Re-attribution
		16. Social support
		16.1 Emotional social support 16.2 Practical social support 16.3 Unspecified social support

Source: Michie, S., et al. (2013). The behavior change technique taxonomy (v1) of 93 hierarchically clustered techniques: Building an international consensus for the reporting of behavior change interventions. *Annals of Behavioral Medicine*, 46(1), 81–95.

Figure VII.1: Taxonomy of behavioral-change techniques (BCTs).
Michie's 2013 taxonomy of 93 behavioral change techniques. Yes, there are so many it is barely legible, this chapter is here to help.

To the change professional schooled in ADKAR or the like, this may seem ridiculously and unnecessarily complicated—however, it is more likely that 93 BCTs are a **simplification** of reality, rather than an over-complication.

How is a change practitioner, with a behavior change imperative, to navigate 93 behavioral change techniques?

One conceptually strong and useful simplification of behavior change is the behavior change wheel (BCW) that offers a systematic approach to understanding behavior in context and provides practical strategies for designing interventions (see Figure VII.2).

BEHAVIOR CHANGE WHEEL

© Michie, van Stralen, & West (2011) • Future of Change Management, Gibbons & Kennedy

Figure VII.2: Behavior Change Wheel (BCW) from Michie's team at University College London (UCL).
BCW helps you work inside out starting with the challenge, then understanding which behavior(s) to focus on, what the barriers and enablers are (drivers) stopping or enabling a behavior to occur, then from there you could select broad intervention types (and policies) and then getting to the nitty gritty of how BCTs can be used to address the drivers.

The BCW links a robust model of behavior change (COM-B, discussed below) with types of intervention and then with policy options. These aren't steps to be followed in a linear fashion but guides to linking behavioral **cause and effect** to behavioral intervention.

We arrive somewhere simpler than 93 techniques, but the change practitioner accustomed to five-step, three-step, or eight-step models may still have a headache. The BCW has 22 boxes—how do we use it?

This chapter offers a method and six tools for using the BCW to help navigate the numerosity of the 93 BCTs.

To do so, we offer a case study:

> Talent Builders has high turnover rates among new employees, which costs a great deal financially. They want to improve the retention rates of new employees within the company during the next 12 months.

How could behavioral science help us help Talent Builders?

Tool 1: Behavioral systems map

One change management tool imported from 1960s research in cybernetics is systems thinking. Our first tool, behavioral systems map, allows its users to identify parts of a system and how they interact with each other to produce a system's output. Take output in a call center, for example. The center's output will be a function of the people (e.g., their skills, attitudes, behaviors, and so forth), and so-called hard structural features (e.g., everything from product quality and pricing to how people are incentivized, to how effectively technology helps them do their job, to performance processes, and much more).

The change questions are: Which of those hundred-or-so performance levers should I pull? Which parts of the system will produce the biggest gains, and in Talent Builders' case, retention?

Behavioral systems mapping allows us to understand components (i.e., behavior and actor, and their drivers, connections, and relationships). It is a useful tool when sense-making, including the technical complexity of different drivers and the social complexity of differing agendas.

Figure VII.3 shows a simplified systems map from our Talent Builders case study.

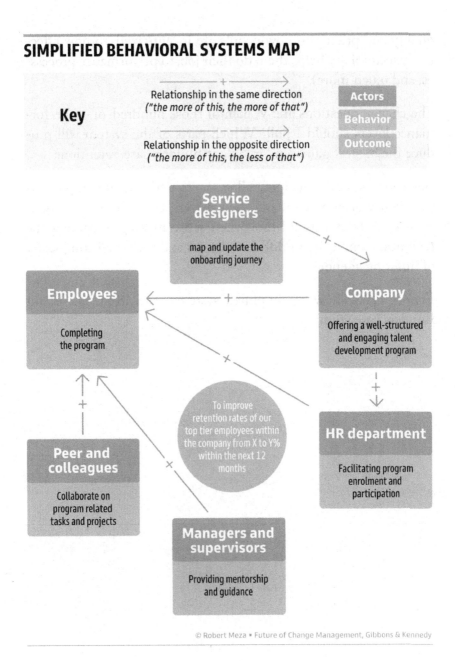

Figure VII.3: A simplified version of a behavioral systems map.
Using a system's map means effort isn't wasted and is directed toward the highest leverage points.

Once we know which aspects of the system provide the desired change with the greatest leverage, we need to descend from that 5,000-metre view and get very specific about which behaviors to target for change.

Tool 2: Behavior specificity

Though most consulting frameworks have a who, what, when, where, how, and why framework, we can direct those questions toward behavioral understanding. Being specific about behaviors helps us avoid unhelpful vagueness, such as 'working more efficiently,' which can mean different things to different people (see Figure VII.4).

BEHAVIOR SPECIFICITY QUESTIONNAIRE

Questions	Use the questions
Who is the target actor?	New employees at the company.
What do they need to do?	Employees to complete the talent development program.
When do they need to do it?	During their first 90 days employment at the company.
Where does it need to happen?	Whitin the company's talent developmentplatform.
How often do they need to do it?	Once a year.
With whom do they need to do it?	With a mentor.
Your statement:	New employees at the company are required to complete the talent development program within the first 90 days of employment. This program should be undertaken once a year, with a mentor and exclusively within the company's designated talent development platform.

© Robert Meza • Future of Change Management, Gibbons & Kennedy

Figure VII.4: Define behaviors as specifically as possible.
Defining behaviors very specifically is an essential early step designing behavioral interventions.

At our fictitious Talent Builders, early completion of their talent development programme "with a mentor" was provisionally seen as providing the highest leverage for improving retention rates.

From this specificity, behavioral scientists next determine root causes of a desired behavior, including barriers to its adoption.

Tool 3: COM-B and behavioral root causes

There are many models of behavior change, but COM-B is a go-to tool for finding root causes of behavior (see Figure VII.5). It is a component of the behavior change wheel we saw earlier, and it introduces three behavioral "macro" dimensions: Capability, opportunity, and motivation (COM).

Capability breaks down further into both physical capability and psychological capability. **Opportunity** breaks down into physical opportunity and social opportunity. **Motivation** breaks down into reflective and automatic motivation.

COM-B COMPONENTS AND DEFINITIONS

COM-B Construct	Definition
Physical capability	Physical skill, strength or stamina
Phychological capability	Knowledge or psychological skill, strength or stamina to engage in the necessary mental processes
Physical opportunity	Opportunity afforded by the environment involving time, resources, locations, cues, physical affordance
Social opportunity	Opportunity afforded by the interpersonal influences, social cues and cultural norms that influence the way that we think about things e.g., the words and concepts that make up our language
Reflective motivation	Reflective processes involving plans (self-conscious intentions) and evaluations (believes about what is good and bad)
Automatic motivation	Automatic processes involving emotional reactions desires (wants and needs), impulses, inhibitiosn, drive states and reflex responses

* Future of Change Management, Gibbons & Kennedy

Figure VII.5: COM-B's components and their definitions.
COM-B analyses the drivers of behavior by taking data and looking for patterns or clusters.

In our case study at Talent Builders, we had selected our target behavior—employees completing an onboarding program—but still need to further understand why this behavior wasn't taking place and what kinds of interventions would help. We deployed COM-B in a workshop format, as illustrated in Figure VII.6.

BEHAVE STRATEGIES

DRIVERS OF BEHAVIOR

From the research we can start bringing over the drivers for the selected target behavior and code them according the COM-B Model.

Challenge:
"Ineffective onboarding processes are leading to high turnover rates among new employees, incurring substantial financial costs for the organization"

Actor:
New employees

Target Behavior::
Employees completing an onboarding program

Outcome:
To improve the retention rates of our top employees within the company from X% to Y% within the next 12 months, thus safeguarding our financial stability and reinforcing our status in the industry

CAPABILITY

Physical Capability:
Physical abilities or condition (strength, stamina) and skills.

Psychological Capability:
Psychological abilities or mental characteristics of people: knowledge, intellectual capacity, memory and decision making processes

Barrier:
a) Employees completing an onboarding program
b) Complexity of or poorly explained program content

Enabler:
Clear instructions how to navigate the program

OPPORTUNITY

Physical Opportunity:
The physical environment with which people interact (objects and events) and the resources available to them (time and money)

Barrier:
High workload and competing priorities

Enabler:
Offering a flexible schedule

Social Capability:
The social environment pople live in: culture, norms, social relationships, and any people who influence their

Barrier:
Employees do not receive social support form managers and peers to complete the program

Enabler:
Having a sense of community / peer collaboration

MOTIVATION

Automatic Motivation:
Emotional reactions, habitual behaviors—things people may not even really understand about themselves

Barrier:
a) Fear of failure or not performing well in the program

b) Lack of intrinsic motivation (self-determination theory). If employees do not find the program intrinsically rewarding or enjoyable, they may lack the automatic motivation to participate

c) Impulse to engage in more immediately gratifying activities, like browsing the internet

Reflective Motivation:
Attitudes, intentions, goals, identity, and values—things that people are able to articulate and often explicitly claim about themselves

Barrier:
Difficulty in maintaining commitment to the program

• Future of Change Management, Gibbons & Kennedy

Figure VII.6: COM-B in workshop format for use with stakeholders.
COM-B is simple and intuitive enough that it can be used in workshops, to gain stakeholder perspectives on the drivers of behavior.

At Talent Builders, the workshop provided the following insights:

- **Psychological capability**: Employees were too overwhelmed by their daily work to engage in an onboarding program (lack of attention).
- **Social opportunity**: Employees did not receive social support from managers and peers to complete an onboarding programme.
- **Reflective motivation**: There was difficulty in maintaining commitment to an onboarding programme.

Still, the COM-B analysis remained too high level for this behavioral change problem, we needed to dig deeper and a COM-B companion tool proves very useful.

Tool 4: Theoretical domains framework (TDF)

The components of COM-B can be further subdivided into 14 domains using the theoretical domains framework (TDF), also from Michie and colleagues. The TDF is an integrative framework that breaks down a single COM-B domain into different kinds of theories to guide intervention. For example, psychological capability might be developed through knowledge, skills, or attentional processes; opportunity might be environmental (such as choice architecture) or social (the effects of group norms or culture).

Figure VII.7 illustrates how TDF domains relate to each COM-B component.

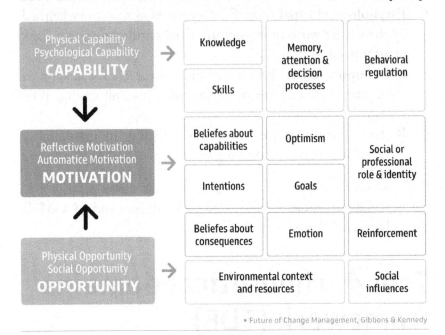

Figure VII.7: COM-B and the Theoretical Domains Framework
The TDF provides a more granular look at the drivers of behavior.

In the case of Talent Builders, we went on to tag relevant TDF domains. This allowed us to develop "solution areas" (such as knowledge) and from there, using the next tool, to consider evidence-based ways of building the required knowledge.

Using TDF, we identified barriers in the "**memory, attention, and decision processes**" (**MADP**) component (such as the ability to retain information, focus selectively on aspects of the environment, and choose between two or more alternatives). Doing this helped us be more specific about how to change **psychological capability**.

Additionally, in the COM-B area of **social opportunity**, we found significant social influences on the desired behavior. In **motiva-**

tion, we identified a barrier around **goals.**

Now we had some candidate levers for changing the intended behaviors at Talent Builders, which led us closer to hypotheses on changing behavior and closer to intervention design.

Contrast this analysis with the behavioral analysis done on a typical change management project. We aren't persuading and praying, or the first three steps of ADKAR, we are targeting specific behaviors. However, even with this more robust analysis, we still aren't quite there.

We want to use **evidence-based approaches.**

For example, if we discover that **social identity** is a critical behavioral change lever, we need to consult the scientific literature on shifting social identity to intervene effectively.

If we discover that **environmental context** is an important lever, we need to consult the scientific literature on which environmental interventions will be most effective.

This can be time-consuming, often too much so for the change practitioner. Few want to wade through dozens and dozens of academic articles to discover which of the techniques they are considering are most robustly grounded in evidence.

Fortunately, there is help for this, too—in the form of the theory and techniques (T&T) tool.

Tool 5: Theory and technique (T&T)

Armed with the TDF outputs, how do we intervene?

We want to use the driver and lever data we analyzed as a guide to find the best evidence-based interventions.

There is an open-source resource from University College London researchers called the theory and techniques (T&T) tool (see full URL in references below[4]) which explores links between BCTs and their drivers. Given the 93 BCTs and myriad of contexts in which they can be used, you'd expect something complicated (see example Figure VII.8).

THEORY AND TECHNIQUES (T&T) TOOL

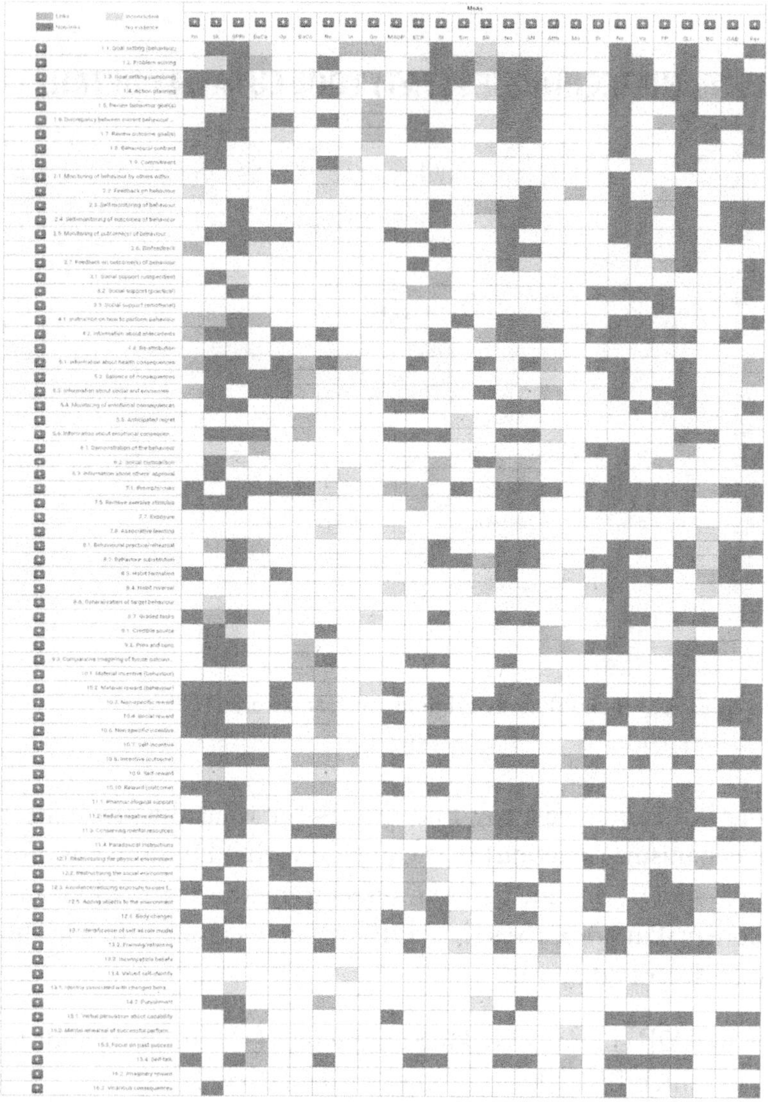

© Robert Meza • Future of Change Management, Gibbons & Kennedy

Figure VII.8: Theory and techniques (T&T) tool in all its complexity.
Once you decide what approach you will take to changing the specific behavior, you want to select an approach supported by evidence.

Figure VII.9 shows a cutaway, closer up version of the sample outputs.

THEORY AND TECHNIQUES (T&T) TOOL CUTAWAY DETAIL

		Kn	Sk	SPRI	BaCa	Op	BaCo	Re	In	Go	MADP
+	5.6. Information about emotional consequen...	■	■		■	■					■
+	6.1. Demonstration of the behaviour										
+	6.2. Social comparison			*							
+	6.3. Information about others' approval							■			
+	7.1. Prompts/cues								*		
+	7.5. Remove aversive stimulus										
+	7.7. Exposure										
+	7.8. Associative learning							■			
+	8.1. Behavioural practice/rehearsal		■								
+	8.2. Behaviour substitution				■						
+	8.3. Habit formation						■				
+	8.4. Habit reversal										■

Legend: Links / Non-links / Inconclusive / No evidence

© Robert Meza • Future of Change Management, Gibbons & Kennedy

Figure VII.9: COM-B and theoretical domains framework (TDF).
We cross-reference an intervention with a TDF area such as skills.

On the horizontal axis, you find drivers (or what the researchers call mechanisms of action) that coincide with TDF domains. For Talent Builders, the final column refers to memory, attention, and decision-making processes (MADP), which were identified in the analyses described above.

In the MADP columns, clicking on the plus (+) button expands to show us which BCTs have a robust evidence base.

The green-colored boxes have the most links to evidence. Yellow boxes suggest moderate evidence, and a blank box indicates no evidence.

In Figure VII.9, look at line 7.1. We see **Prompts/cues** in green so there are a greater number of links to evidence. In contrast, line 6.1 **Demonstration of behavior** has no evidence. We may still choose a tool for which there is little evidence, but we are in virgin territory; often, we will want to start where there is the most evidence.

Once organizations become adept at using behavioral science, they should consider building their own evidence base after they test these techniques to make sure they are not only using the science and evidence from the literature but also their own practical and context-specific data.

At Talent Builders, we focused on line 7.1 (**Prompts/cues**). Then we prototyped two kinds of interventions: Digital reminders and visual timetables (see Figure VII.10). This stage is where art meets science. We use creativity and imagination to design interventions identified with science.

BEHAVIORAL CHANGE TECHNIQUES (BCTs)

Concept	Detailed Explanation / Concept	Steps Needed
1. Peer buddy system	Pairing new employees with experienced ones, allowing for a smoother transition, mutual learning, and inmediate social integration.	1. Identify experienced volunteers. 2. Set guidelines for the pairing system. 3. Match based on role/ interest. 4. Monitor & gather feedback for continuous improvement.
2. Group onboarding sessions	Pairing new employees with experienced ones, allowing for a smoother transition, mutual learning, and inmediate social integration.	1. Identify group tasks or learning modules. 2. Schedule sessions in advance. 3. Facilitate with experienced staff or trainers. 4. Encourage group discussions and activities.
3. Open forum discussions	A platform where all employees, new or experienced, can discuss challenges, share experiences, and offer support.	1. Choose a platform (e.g., intranet, chat room). 2. Promote its existence and purpose. 3. Monitor for constructive discourse. 4. Periodically hightlight valuable discussions.

© Robert Meza • Future of Change Management, Gibbons & Kennedy

Figure VII.10: Final output of the T&T tool.
Using the T&T tool allowed us to design and prototype two kinds of interventions.

Finally, we have practical concerns – is it acceptable, affordable, and fair, for example? Our final tool helps assess such practical concerns.

Tool 6: APEASE framework

Before scaling our interventions, we should evaluate our solution along a number of practical criteria: Acceptability, practicability, effectiveness, affordability, side effects, and equity (the components of the APEASE framework). APEASE is used to assess people-related policies in business. Are they fair? Are there side effects? APEASE applies some useful, pragmatic tests (see Figure VII.11).

APEASE FRAMEWORK

© Michie, Atkins, & West (2014) • Future of Change Management, Gibbons & Kennedy

Figure VII.11: APEASE tests for practicality.
Simple checklists like APEASE are useful for making solutions fit.

At Talent Builders, we used APEASE to slow down our gut feelings about which concepts we got most excited about. Instead we asked teams to score our solutions along these APPEASE criteria. The results of APEASE analysis at Talent Builders are captured in Figure VII.11.

USING APEASE TO ASSESS PRACTICALITY

Affordability:	No direct costs, but manager's time might be a factor.
Practicability:	Requires buy-in and communication.
Effectiveness:	Likely high, as managerial support can significantly influence employee priorities.
Acceptability:	Generally high, as long as it doesn't add undue pressure.
Side effects:	Minimal, though there's potential for stress if not approached correctly.
Equity:	Fair if all employees receive consistent support.

© Robert Meza • Future of Change Management, Gibbons & Kennedy

Figure VII.12: APEASE allowed Talent Builders to examine many implementation issues often ignored.
We used APEASE in workshop format at Talent Builders to great effect.

Conclusion

Human behavior is complex. In just the last decade, our understanding of it has vaulted forward and our approach has become more systematic. Bearing in mind behavior's inherent complexity, we need to analyse behaviors for context and use practical tools to affect this unavoidable complexity into something more manageable for the change practitioner to intervene.

The tools demonstrated and illustrated hopefully pass the Goldilocks-Einstein test, or simple but not simplistic. They are grounded in the latest science and are often supported by a substantial evidence base, and also allow a change practitioner to think more creatively and exhaustively about behavioral change and how to affect it for their clients. This, we hope, avoids the more common knee-jerk interventions—and such as 'just throw them a bit of training'— and considers the plethora of systemic factors that affect behaviors in organizations.

References

[1] Michie, S., Atkins, L., & West, R. (2014) *The behaviour change wheel—a guide to designing interventions*. Silverback.

[2] Michie, S., et al. (2013). The behavior change technique taxonomy (v1) of 93 hierarchically clustered techniques: Building an international consensus for the reporting of behavior change interventions. *Annals of Behavioral Medicine, 46*(1), 81–95.

[3] Michie, S., van Stralen, M.M. & West, R. (2011). The behaviour change wheel: A new method for characterising and designing behaviour change interventions. *Implementation Science, 6*, 42.

[4] https://theoryandtechniquetool.humanbehaviourchange.org/tool

Further Reading

Meza, R. (n.d.). *Behavior design tools*. https://courses.aimforbehavior.com/free-behavior-and-innovation-frameworks

CHAPTER 8

Uses and Abuses of Design Thinking

by Yves van Durme and Paul Gibbons

Design thinking has stormed the ramparts in the last decade and now is used throughout leading businesses in areas as diverse as strategy and organisation design. The design mindset solves some of the problems change leaders have historically faced, such as engaging stakeholders, developing creative solutions, and organisational learning. In some circles, though, it is seen as a panacea and a substitute for change management. Paul and Yves think this an error, point out the pitfalls, and argue that change management and design thinking go hand in hand.

Death of mushroom change management

Sometimes, companies commit change suicide. Only 10 years ago, we saw the stubborn ignorance of change principles torpedo multi-million-dollar projects with application of mushroom change management. "Keep them in the dark and feed them s$#t." Sounds pithy. Clickbait. Do any businesses do that during major change?

Back in 2008, the CFO of a major airline declared, "We wouldn't ask the engineers what their views on our software systems (for engineers) were. We'll put in what we think is appropriate for us." **Keep them in the dark.**

Ten years ago, the C-suite's attitude toward change was sometimes hostile. BP's global head of HR said of the change management plan for a $100 million project: "I certainly do not want my people sitting in beanbag chairs, next to lava lamps, talking about their feelings during this project. Tell them what to do and have them bloody do it." **Feed them s$#t.**

Both projects failed and both execs were turfed out. A cognitive bias, the Dunning-Kruger effect, predicts an inverse relationship between confidence and knowledge (you read that correctly). People who know the least can be the most dogmatically confident in their views (take climate change and vaccines).

Dunning-Kruger killed change at those companies. The chief people leader (some irony, huh?) and CFO both failed to understand that change management isn't about yoga and patchouli, it is about people producing results with each other, for each other, and through each other.

Since those execs collected their pink slips, we have come a long way. Lamentably, the mushroom approach lingers on. "Build it, then sell it" is the paradigm. The folks in charge decide what needs to happen, and the folks at the bottom are cajoled, persuaded, influenced, coerced, swayed, persuaded, guided, inspired, educated, or threatened into compliance (see figure VIII.1 for a "client" case study).

AFTER-THE-FACT INFLUENCE OFTEN INVOLVES COERCION

Figure VIII.1: Influencing stakeholders after the fact can require coercion, sometimes extreme.
In Game of Thrones, Queen Danaerys used Drogon to influence key stakeholders in her preferred direction.

Today, we know better. Persuading and inspiring people previously left out of the process is fraught with difficulty—at best, with luck, it produces a tepid sort of enthusiasm. People do not like to be engaged as an afterthought. The effectiveness of the "build, then sell" paradigm was poor, and the failure rates for significant change programs reflected that.

Companies know better now. However, as the maxim goes, "a little knowledge can be a dangerous thing." One change risk today comes wrapped in an attractive package—design thinking. How could design thinking, a powerful tool, possibly be a threat to major change programmes?

Designers do it better—creativity, empathy, engagement, and learning from the outset

"Empathy is the skill of the future, and practicing empathy every day as a business leader helps you understand what your customers and team need."

FREDERIK G. PFERDT (GOOGLE'S CHIEF INNOVATION EVANGELIST)

Enter design thinking (see Figure VIII.2). As with change management, design thinking starts with people: people producing results with each other, for each other, and through each other. Design thinking builds in engagement, creativity, and learning **during** the solution design process. This has always been the aim of sound change management but has remained a seldom realized ideal. More often, change teams were summoned once the solution had been selected, or worse, once the project went off the rails. Indeed, could we wave a magic wand, we would have clients start their change process, engaging stakeholders, long **before** they write the request for proposal, long before they select their consultants, and long before they start the project!

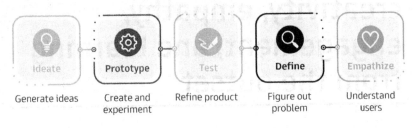

Figure VIII.2: Design thinking process.
Design thinking is an iterative process that builds in creativity, empathy, engagement, and learning from the get-go.

Empathy, the first stage in many design thinking models, is a step not found in any brand name, prescriptive change models such as Kotter's eight steps, Beer's six steps, Prosci's ADKAR, Kanter's ten commandments, or Lewin's unfreeze. There is vague stuff about guiding coalitions and communications, but in all those models, at least on the surface, communications are one way: Ain't no connectin' with ya people! The leader's job is broadcasting, sometimes listening, rarely empathizing.

With design thinking, solution designers walk a mile in the shoes of "users," be they employees or customers, engaging with them emotionally as well as around their needs. In this way, design thinking has advanced the practice of change management beyond its 20th-century roots.

You would think that all the change gurus of yesteryear would have stuck some creativity someplace in their models. Nope. Creativity is not found in those "branded" change models, usually not even mentioned, and certainly not a step in the process. To be sure, some creativity happens because, you know, human beings. But it happens **by accident and not by design.** This extraordinary omission means even the highly structured change management, decision making, and strategy processes of bygone days, bizarrely, have never had **creativity hard-wired** into them!

Creativity sometimes entered the change process through the skills of facilitators, but those expert facilitators of strategy, change, and decision-making processes had just a few tools, mostly brainstorming. Recent research suggests this the go-to creativity tool, brainstorming, may be an unsatisfactory method of introducing divergent thinking despite its ubiquity.

The creativity toolkit used in design thinking, such as rapid prototyping, storyboarding, SCAMPER, personas, customer experience journeys, criterion matrices, POINT, paired comparison analysis, and brainwriting, were outside the skillset of most traditionally trained facilitators. These empathetic and creative processes give users skin-in-the-game, a stake in the solution's success. They release the passion teams can generate when they create novel, valuable solutions together.

Design thinking also, because of its high engagement methods, begins to solve another age-old change management problem: **Skills transfer**, that is, teaching users to use the new system, process, or product. Historically, in a mushroom change management approach, we would underinvest in user engagement and skills.

Or, when we took it seriously, we would find the learning curve too steep because **learning requires** engagement, and mushrooms, when they emerge from darkness, aren't engaged at all.

With design thinking, learning, building skills, creating new knowledge, working cross-functionally, becoming engaged, and solving problems happen concurrently, bringing us closer to the holy grail of learning 2.0—a nut that few businesses have wholly cracked.

Design thinking is the new black

> *"Design thinking can be the Trojan horse inside which empathy and engagement—key change management outcomes—are hidden!"*
>
> **PAUL GIBBONS**

So powerful is design thinking as a tool that it is no longer used just for **product design,** as was primarily the case a decade ago. Now it is also used to design digital and physical customer and **user experiences**—so much so that we have called this the experience revolution. Design thinking is increasingly recognised by the C-suite and has even been used to **formulate strategies,** to **redesign workflows,** and to **transform business models** (see Figure VIII.3).

HIGH-ENGAGEMENT DESIGN AND PROBLEM SOLVING

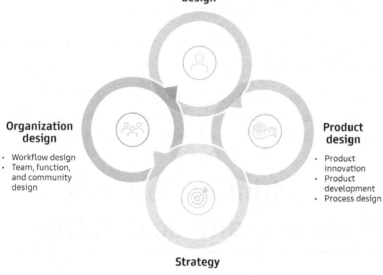

Figure VIII.3: Where can we use the design mindset in business?
From its cool-kids roots, design thinking has become the go-to tool for high engagement design and problem-solving processes.

In all those areas, design thinking has produced prodigious results, results that could not otherwise have been easily obtained. For example, one of our pharmaceutical clients faced a precipitous drop in revenue due to patent expiration. Nevertheless, R&D and clinical trial costs, and the time involved in drug design, were skyrocketing. This persuaded them to reinvent their traditional development process. Using a "fast track" high-engagement process with stakeholders (i.e., customers, physicians, employees, and reps) company leaders came up with ideas for new markets, products, business models, and pipelines. This immersive environment generated not just a wealth of **creative ideas**, but as importantly, the opportunity for senior leaders to **connect deeply** with stakeholders in what they called a unifying experience.

So successful is design thinking that we believe it risks being a victim of its own success.

We fear, with some evidence, that teams without the change management battle scars may see it as a substitute for sound change management and not just a tool that works best in harness with it.

Design thinking pitfalls—alignment and scalability

Social science research from the 1940s showed that asking users about a change makes it more likely they will align with and adopt it. Although we waited 70 years for that insight to take root—design thinking helps us magnificently. The empathy, engagement, creativity, and prototype testing produce substantial breakthroughs in alignment, much more than the mushrooms and "build-it then sell-it" paradigms.

The new risk we see is that design thinking is seen as a **change panacea**. In other words, design thinking grants teams an "excused absence" from change worries such as stakeholders, politics, culture, and resistance. We have seen many executives pop the champagne cork too soon thinking, "We did the whole alignment thing already."

This misapprehension is understandable because design thinking builds in critical empathy and engagement processes **from the start** (i.e., it has some change management built in). However, alignment, even of users deeply engaged in the design process, isn't a "one and done." Humans just don't work that way, and **constant communication and involvement alignment "re-ups"** are essential, even among stakeholders deeply involved in the design process.

Worse though, some stakeholders may not have been involved. Sure, we would like to involve everybody. However, the realpolitik of change and time-and-money constraints may mean some stakeholders are left out. Not only do those excluded stakeholders **start less engaged**, they may also **feel antipathy** from having been excluded. (See Figure VIII.4 for a case study in disenfranchised stakeholders.)

FRENCH REVOLUTIONARIES STORM THE BASTILLE

Figure VIII.4: Parisian stakeholders storm the Bastille.
Involve stakeholders or risk lack of alignment and even antipathy toward project goals. They might even storm your castle (Prise de la Bastille, Anon, 1790).

This problem becomes particularly acute as we try to **scale prototypes** that emerge from a design thinking process. A successful prototype, even a wildly successful one, may encounter the NIH problem, or "not invented here." Indeed, we have seen a successful culture change prototype project get pilloried by the rest of the business—those inspired and empowered individuals, proud avatars of the new culture, so pumped by the design thinking process used to design the culture change, appear (to the rest of the business) to have "drunk the Kool-Aid."

From a business perspective, being **first to scale, not first to design**, matters most. Change management is the transmission that connects the design engine to the implementation drive train. Businesses cannot get full-scale adoption without the messy business of working through commitment, political, and cultural issues.

Too often, we've seen terrific prototypes "thrown over the wall" to the build team. The build team hasn't the customer intimacy nor the passion of the design team, and sometimes the novelty of the solution will be dulled by force-fitting it to legacy systems. These multiple handoffs are where tons of value can "leak" from the design process.

We have seen our clients deploy design thinking but forget more traditional change management interventions: Training, sponsorship development, communications strategy, and risk management. It is easy to forget that new ideas and new behaviours compete for attention in a noisy world.

One solution is to think of the team's outputs as part of a "campaign" from the outset, and our job is to get clicks, likes, subscribes, and shares from beyond just the design thinking team—what is sometimes called a "show your work" mindset. To that end, our savviest clients have found enterprise social networks, such as Slack and Teams, of great value: Stakeholder management, communication, problem-solving, creativity, knowledge management, learning, and engagement can happen in concert.

More darkly, when we have seen this "campaign" ignored, all the astonishing creativity, engagement, empathy, and learning which was inspiring to behold, fell afoul of the same **cultural, political, and systemic** issues that stymied solutions developed by traditional means.

DESIGN THINKING, LEAN STARTUP, AND AGILE INCREMENTALISM

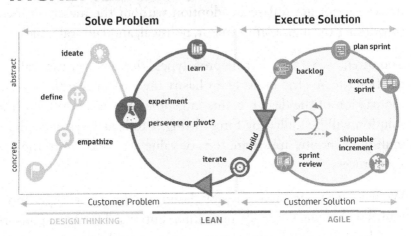

Figure VIII.5: Figure title. Design thinking, Lean startup, and Agile diagram. Full stack designers can integrate design thinking, Lean, and Agile, but must use change management to drive the entire process.

Our tech colleagues talk about full-stack developers, and super-technologists who can deliver the whole technology stack, from website and UX design, through databases, servers, DevOps, APIs, and mobile. Our goal is to offer full-stack designers who can shepherd a design process all the way to scaled implementation, perhaps using sister methodologies such as Lean startup and Agile (see Figure VIII.5). However, full-stack designers must, above all, be change management experts, for those are the skills necessary to help a client navigate interpersonal, cultural, systemic, and political barriers to implementation.

Integration—key to unlocking value

We have come a long way since cookie-cutter approaches to change from the 1990s. Design thinking has added an indispensable tool to the change (or strategy) consultant's toolkit. Executives see its value through the prodigious creative outcomes it produces.

However, using design thinking does not provide an all-access backstage pass beyond the business' power structures. It helps with alignment in a limited way and scaling even less. We encourage our clients not to think of design thinking as a change panacea, but rather as part of an **integrated approach to change**. Only in that way will we realise design thinking's full potential.

References

[1] Stroebe, W., & Diehl, M. (1994). Why groups are less effective than their members: on productivity losses in idea generating groups. *European Review of Social Psychology, 1*(5), 271-303.

Further Reading

Brown, T. (2008, June). *Design thinking*. Harvard Business Review.

DeBellis, P. (2020, December 2). *Adaptable by design: a future-focused, fit-for-purpose HR operating model*. Deloitte. https://www2.deloitte.com/us/en/blog/human-capital-blog/2020/hr-operating-model.html

Mazor, A., Johnsen, G., Stephane, J., Hill, A., Calamai, J.B., & Moen, B. (2019). *Exponential HR: break away from traditional operating models to achieve work outcomes*. Deloitte.

Solow, M., & Wakefield, N. (2016, February 29). *Design thinking: crafting the employee experience*. Deloitte. https://www2.deloitte.com/us/en/insights/focus/human-capital-trends/2016/employee-experience-management-design-thinking.html

CHAPTER 9

Will ChatGPT Replace the Change Manager?

by Natasha Young

ChatGPT burst onto the scene in late 2022 and inventive change practitioners have begun to ask, 'How can this help my clients and I implement change?' Natasha is a work-a-day change management expert at IBM whose days are filled with helping clients navigate complex change problems, and she is an early generative AI adopter. In this chapter, she shares her experiments with ChatGPT within her change practice.

Rerum novarum cupidus (greedy for new things, Cicero)

IBM CEO Arvind Krishna calls it a 'Netscape moment,' as groundbreaking as the first internet browser 30 years ago. Alphabet (Google) CEO Sundar Pichai calls it 'more profound than fire and electricity.'

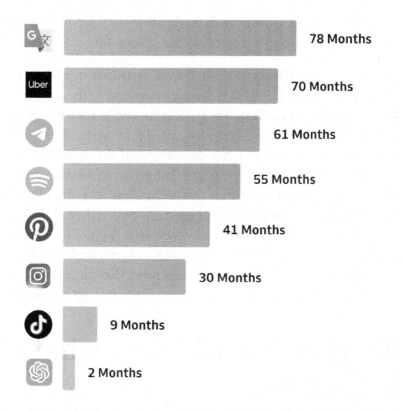

Figure IX.1: ChatGPT took only months to achieve what other 'exponential' apps did in years.
Even faster than TikTok? It took Uber six years to achieve what took ChatGPT only two months (data from Statista).

The 'it' these CEOs are talking about is, of course, generative artificial intelligence (GenAI for short), or advanced machine learning that allows computers to generate contextually relevant information, content, art, writing, music, and computer code, to name but a few.

ChatGPT, where GPT stands for generative pre-trained transformer, is the most widely known of these applications. Figure IX.1 compares its user growth with earlier exponential applications. Perhaps Cicero was right, we are greedy for new things.

GenAI is a watershed moment for AI (artificial intelligence) in two primary ways. Firstly, it is raising awareness of the potential of AI among the public. It makes it clear to a wide audience that we have moved beyond the Turing test of 70 years ago and Deep Blue of 25 years ago. ChatGPT has demonstrated the ability to generate creative and nuanced responses that go beyond mere parroting of pre-programmed responses. Secondly, ChatGPT has put AI in the hands of the masses. The biggest allure of ChatGPT is that it is easy to use. It provides wide access to a simple interface that produces answers that are, in most cases, not only well-constructed and believable but also adopt a human tone. This surge in popularity is pressuring businesses to get in on the act or get left behind.

What is ChatGPT and what is it used for?

ChatGPT uses advanced natural language processing and machine learning algorithms to enable human-like communication between users and digital assistants. By analysing vast amounts of data, it can generate insightful and personalised responses to a wide range of queries, from providing information and answering questions to making recommendations, and even engaging in creative writing.

As such, ChatGPT represents a major step forward in the development of intelligent digital assistants that interact with humans in more intuitive, natural, and productive ways. It is constantly learning and adapting and is revolutionising every facet of work and workplaces.

Now what if I told you that the paragraph above was written by ChatGPT itself? Clever, right? Not perfect but overwhelmingly impressive.

A few emerging use-case categories are shown in Figure IX.2.

Figure IX.2: GenAI use cases are proliferating, a few examples from just six months after launch.
But can it cook breakfast? As the world tinkers with GenAI, use cases continue to expand.

If you believe the punditry, we are just getting started and, as of this writing, ChatGPT is only one year old!

A report by Goldman Sachs suggests AI could replace the equivalent of 300-million, full-time jobs, as well as a quarter of work tasks in the U.S. and Europe. However, panic not, the same report also predicts that like previous industrial revolutions there will be new jobs and a productivity boom. Additionally, an early impact study of GPTs on labour-market coding released by Cornell University found that approximately 80% of the U.S. workforce could have at least 10% of their work tasks affected by their introduction, while around 19% of workers may see at least 50% of their tasks impacted. On the plus side, professionals can use generative AI technologies to their advantage to improve their work, increasing their productivity and allowing them to focus on higher-value tasks. Research has found that customer service workers at a Fortune-500 software firm who used generative AI tools became 14% more productive than those who did not, with the least-skilled workers obtaining the most benefit.

ChatGPT and change management

I previously experimented with ChatGPT for fun, asking it questions such as, 'How would Shakespeare have said this?' or 'How would you turn this into a haiku?' Then, a few months ago, I started a new programme of work as its change manager. I then started an experiment with questions like "How can ChatGPT help me, my clients, and IBM deliver complex change?"

Here are some of the organizational change use cases that our team has tried (see Figure IX.3). I can't share our client-specific data but can illustrate in general terms what worked and what did not.

ChatGPT AND CHANGE MANAGEMENT — FIRST EXPERIMENTAL USES

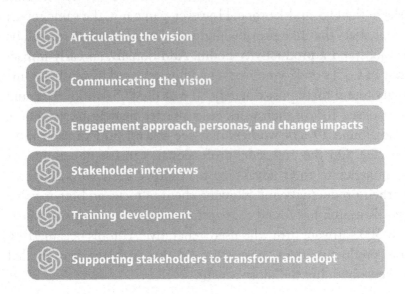

Figure IX.3: Organisational change use cases for generative AI (e.g., ChatGPT).
It is early days for ChatGPT and organizational change, so far, it has proven useful.

Articulate the vision

Even the most complex change needs a concise, persuasive, and inspirational vision. A vision statement will often link market trends to business strategy to a change programme: The why, the what, and the how.

ChatGPT generated the following vision for a cloud computing implementation:

> By migrating to the cloud, you can shed the constraints of traditional on-premises infrastructure, significantly reducing upfront hardware costs and ongoing maintenance expenses. Cloud services provide the scalability to accommodate your evolving needs, ensuring that you can swiftly adapt to market changes and seize opportunities without the burden of infrastructure management. With the added benefits of accessibility, robust security measures, automatic updates, and disaster recovery, cloud computing empowers your organization to focus on core business strategies, innovate faster, and remain agile in a rapidly evolving digital landscape.

That rendition is dry (as dirt), but it does cover the bases. Moreover, we can quickly generate five, ten, or twenty of these playing with different prompts, possibilities, scenarios, and constraints.

Humans re-enter the system to select between generated options, perhaps combining the best of several in a workshop format.

Then, of course, the humans do what (for now) only they can—add emotion, inspiration, and motivation, as well as context, to color the output and to make it warmer, or less dry.

Communicate the vision

As a change consultant, I often write and circulate communications across a company. Sometimes that means communicating the vision to build engagement; sometimes it is sharing information about changes and their effects.

Imagine my delight when I realised that ChatGPT could develop newsletters, information packs, and tailored communications at the click of a button! Indeed, ChatGPT turns out to be a very good writer. It can pretend to be Shakespeare or Yeats; some users have used it to write essays or newspaper articles, and even dating site enticements.

What I found most helpful were its ability to:

a) Produce early, strawman drafts quickly,
b) Summarize text and produce one-page descriptions, and
c) Tailor a message to different constituents.

I even asked ChatGPT to offer me metaphors or analogies when I thought they might help.

What I learnt was the importance of being clear about what I wanted to produce, and providing the context, including the purpose, audience, and tone—such as professional, persuasive, engaging, conversational, informative, authoritative. This is now a core GenAI meta-skill called **prompt engineering**.

It was important to iterate with ChatGPT as I produced communications, to sharpen, refine, and improve results. For a human to produce multiple (dozens) of versions would be time-consuming—ChatGPT does it instantly. Different versions can then be A/B tested for effectiveness and reach.

Personas and change impacts

In the context of managing organizational change, a persona can be defined as a discovery-based representation of different stakeholder groups impacted by a transformation. The philosophy behind creating personas is simply walking in the shoes of your potential user or stakeholder, helping to understand their perspective, needs, motivations, pain points, and more.

Developing personas is usually a lengthy process, characterised by end-user workshops, in-depth analysis, iteration, and refining. I realised this could be made vastly more efficient with ChatGPT. Nevertheless, to create a persona that would respond as desired, there were still key factors I needed to inculcate in my prompts:

- Choosing the character the persona would represent—an end-user or stakeholder group affected by the change.
- Shaping the persona into a character. To do this, I defined the behavior, likes and dislikes, values, goals, and aspirations of the persona. Then I would add status, location, role, and emotions. These traits are necessary to develop a character that can give a human-like response.
- Setting the range of experiences the persona had experienced, thereby giving a depth of knowledge (e.g., the response maturity and demands will vary considerably between a sales manager with 20 years of experience and a recent graduate).
- Specifying the severity with which the persona acts in various situations and fine-tuning the emotions keeps the persona in its shoes.
- Giving personas a scenario and specific goals to imagine their responses.

By asking the personas questions, I was able to generate insights for gathering change requirements and impacts. It was still important to test and mold the responses given by checking for the accuracy of its answers, noting the points where it could act better, and then including those in the next prompt.

Stakeholder interviews

Though some 'temperature taking' of stakeholders happens by survey, more granular feedback comes from interviews and sometimes focus groups. These interviews may generate, for example, 50 or so 30-minute transcripts—in the region of 500 pages of text.

Although rich with data, qualitative analysis requires thematic coding—that is labeling and categorising themes that emerge from the data—a laborious process.

With ChatGPT it is now possible to get the same level of rich output from qualitative data in the amount of time that would be spent on the analysis of quantitative data.

The process I used in this experiment involved:

- Recording interviews with stakeholders and even dry running those interviews with a ChatGPT persona.
- Transcribing the interviews with a transcription app, such as Whisper.
- Writing prompts to specify the insights I want to generate from the interviews (e.g., a summary, issues and blockers, or follow-up meetings to schedule).
- Generating meeting transcripts and insights using GPT algorithms.

To take this a step further, ChatGPT and sentiment analysis then helped me to understand how stakeholders might feel about a change vision and identify potential areas of resistance or concern. I used it to solicit feedback and ideas, which identified potential issues and opportunities that may have been overlooked, in turn, increasing engagement and ownership in the change process.

Additionally, I used ChatGPT's advanced data analytics, including predictive modeling and natural language processing abilities, to analyze employee feedback and performance metrics and identify areas where further improvements could be made. This provided insights that further helped to refine the change process, improved the likelihood of success, and addressed potential concerns or objections before they became a blocker to change.

Training and development

Change management often has a training component for which ChatGPT proves extremely useful.

First, I developed a training-needs analysis by revisiting the personas I created. ChatGPT served as an intelligent tool for identifying the skills and competencies that were required for job roles within the project scope. By processing large amounts of data and applying its advanced understanding of language, it was able to provide comprehensive lists of skills required for the various roles.

> **Example prompt**: 'What are the essential [technical, soft, management] skills and [industry-specific, transferable] competencies required for [specific job title or industry] in [specific location or global market]?'

Second, ChatGPT was able to assist with comparing skills and competencies of the current employees against identified project requirements. By analyzing responses or text inputs, it could identify areas where the end-user groups may be lacking, thereby providing a clear picture of skill gaps.

Finally, once gaps were identified, ChatGPT was able to recommend areas for training and development. It even can suggest specific training programs or courses, thereby helping to bridge any skill gap effectively. This worked to reduce the learning curve and improve the success of the change initiative.

The part I found more useful than perhaps any of the above, was that by inputting specific training requirements and parameters, such as target audience and learning objectives, ChatGPT was able to tailor training to individual user needs, allowing users to learn at their own pace and in a modality that they might prefer. To do this, ChatGPT analysed the input information and generated human-like tailored responses such as explanations, examples, and instructions. I incorporated this tailored output into the final training materials to help employees adapt to new processes and systems.

Stakeholder support

As well as training on the new processes, end-users in a change programme also need a level of personalized support throughout the initiative. I fed information about the programme into ChatGPT and gave this retrained chat model to end users. ChatGPT was then able to answer questions and provide resources to help employees feel more comfortable and informed during the change process. As ChatGPT is accessible from anywhere, it was thought of as a convenient tool for those end-users who were working remotely or had more demanding schedules.

What ChatGPT taught me

With the emergence of generative AI such as ChatGPT, it is apparent that the professional practice of change management is on the cusp of its own major transformation. Thought leaders predict that generative AI will enable companies to make faster, more data-driven decisions and execute change initiatives with greater speed and precision than ever before.

Consulting has always been a 'people' process. However, as technology and AI rapidly develop, the gap between traditional face-to-face consulting styles and modern client needs has never been more pressing. Modernising the consulting process will likely involve utilising AI tools in the workplace, including ChatGPT, as something of a data analyst. As organisations transition to a greater use of AI, change management professionals, like me, need to adapt to the benefits and challenges that AI presents.

Is the professional change manager obsolete?

Does this mean I suggest that change managers are obsolete and that their roles should be automated using ChatGPT? Not at all. Despite its convincing abilities, ChatGPT has limitations.

Lacks emotional intelligence

In the context of change management, this can make it difficult for the tool to accurately interpret and respond to emotional cues from employees. This can limit its effectiveness in addressing employee concerns and fostering engagement in the change management process, which is critical to success. Importantly, change management is human-centered. It is currently impossible for an AI tool like ChatGPT to fill this gap. It lacks the necessary human connection. It is devoid of 'human' skills like the ability to understand and influence people, to read the room, to manage conflict, to create genuine connections, and to tailor strategies and tactics to specific contexts. While it can provide valuable support and information, it cannot replace the importance of human interactions and relationships in driving successful change. Perhaps it acts as a change manager's assistant, but at this stage, it will not replace them.

Limited contextual understanding

It can understand natural language inputs, but it may not always understand the context of a conversation or the nuances of a particular situation, especially in regard to organisational change.

Importantly, individual organisational norms and values are not reflected in responses and, as such, organisations will need to adapt the technology to incorporate their own culture and values—an exercise that requires technical expertise and computing power beyond what most change managers have access to. This limited contextual understanding makes it challenging for ChatGPT to provide personalised or customised responses to employees as could be done with a response from a human change manager. In turn, this can limit its effectiveness in driving change.

Strong dependence on data

Currently, foundation models do not get live feeds from the internet, meaning that their knowledge and understanding of current events or topics are limited to the scope and currency of their training data. This limits their usefulness in providing real-time support or information to employees during a change initiative. Moreover, while ChatGPT can draw on a deep well of knowledge and understanding to provide helpful insights and recommendations, its responses are based solely on the data it has been trained on and the rules behind it. For example, as ChatGPT is not a fully-fledged change impact assessment tool, its responses should be taken as a starting point rather than a definitive source of information. Ultimately, a human expert with experience in conducting change impact assessments needs consulting to ensure relevant factors are considered.

Provides lowest common denominator knowledge

One question I asked was whether ChatGPT 'knew' about change management—whether its abstract knowledge of key concepts was robust. It was familiar with some older (editorial comment: read obsolete) ideas, such as Kotter's eight steps and ADKAR.

However, it still lacked practical knowledge of how to apply such knowledge. (see Figure IX.4).

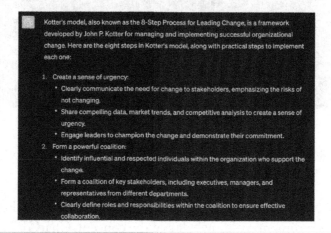

Figure IX.4: ChatGPT doesn't evaluate validity, only popularity.
If you used ChatGPT in 1400, it would tell you that the world is flat; it is limited by the veracity of the internet.

I asked specific questions such as, 'What are the five most common barriers to effective change?' It had a naïve view of resistance, which is a much more complex phenomenon than lack of understanding, or anxiety about roles (see Figure IX.5).

NAÏVE VIEW OF CHANGE RESISTANCE

> Implementing change within an organization can be challenging due to various barriers that hinder its effectiveness. Here are five common barriers to effective change:
>
> 1. Resistance to change:
>   Resistance is a natural reaction to change and can stem from fear, uncertainty, or a perceived threat to job security or existing ways of doing things. People may resist change due to a lack of understanding, concerns about the impact on their roles, or a general aversion to change. Overcoming resistance requires effective communication, involvement of key stakeholders, addressing concerns, and providing support and training.

Figure IX.5: ChatGPT has naïve view of change resistance.
With its concise answers, ChatGPT risks oversimplifying.

I then asked for 'talking points' on resistance to change. Here, despite its rudimentary knowledge of the concept, ChatGPT provided a useful checklist for working with resistance (see Figure IX.6). Moreover, although ChatGPT cannot get me out of a meeting with a difficult stakeholder, I can use it to role-play the conversation, giving me a head start.

USEFUL LIST OF RESISTANCE TALKING POINTS

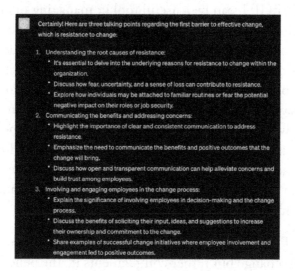

Figure IX.6: Useful list of talking points returned when ChatGPT is asked to role play a conversation about change resistance.
The change manager can role play tricky conversations with stakeholders.

These answers remind me of a general concern that we ought to worry about, both as humans and change professionals: ChatGPT provides answers that are average and/or pedestrian. They might be, as with traditional search, the most popular answers; but also as with traditional search, it is difficult to weigh expertise and excellence. Google, despite its sophistication, still includes 'page rank' as a signal in ordering its responses.

As with all things internet, 'clicks' become a proxy for credibility, and 'popularity' becomes a proxy for expertise. Using ChatGPT well depends on knowing when to question or discount its responses, tempering its answers with human judgments that it today cannot match.

Can change managers and ChatGPT coexist?

While ChatGPT can be a useful tool in managing organizational change, the associated ethical considerations, risks, and challenges indicate that it should be used with other strategies and methods that consider its limitations. After all, ChatGPT is just a tool and cannot replace the human element of change management. In this way, it may disrupt the execution but not the ideation of change. In other words, AI disrupts the execution of tasks and activities, but that just makes the concepts and ingenuity behind those tasks more important. ChatGPT helps draft a project-wide communication, for example, but still requires the human change professional to input the right ideas.

ChatGPT does not replace the human touch of managing organizational change but rather complements it. The benefit here is it frees up time for the human change managers to concentrate on developing and using skills that machines do not yet have. These include strategizing, innovating, complex problem-solving, and demonstrating emotional intelligence and empathy. Combining ChatGPT with effective leadership, communication, and collaboration can maximise benefits to the business and drive successful change initiatives. It is also an opportunity to hone human skills likely to become increasingly important for career development as we progress into the era of AI and automation. This is an excellent example of where humans and machines working together can achieve more than machines or humans working separately. Thought leaders in this area, including Professor Thomas Malone of MIT, have advocated for a collective intelligence approach: "People and computers can be connected so that collectively they act more intelligently than any one person, group, or computer has acted before."

As such, perhaps the more relevant discussion than how generative AI will lead to job displacement is how it will create new opportunities for workers with the skills to manage and work alongside machines. Instead of feeling threatened by technological advancements, including generative AI, how can we better implement technology to support workers and improve jobs?

How will generative AI change our lives and the way we work?

The big question remains: How will generative AI change our lives and the way we work as change management professionals? The simple answer is that nobody knows. There are many different opinions about generative AI; however, the truth is that technology does not care about our opinions and our opinions will not stop its advance. So, we too should advance in line with it. If AI has the capability to take over part of your role, then you can either prepare yourself for change or equip yourself to use it as a tool.

> ***Editorial comment***: There are valid concerns amongst experts about some aspects of generative AI's introduction to society and workplaces, especially as its capabilities grow and expand. Within the context of this chapter, or its application to organizational change, we believe fear of role replacement is overrated compared to the potential benefits of assisting change managers.

Nevertheless, humans maintain control. Generative AI creates efficiency, and indeed, makes some jobs less valuable. But that has been the story of humankind—we evolve, we adjust, and we find new opportunities. If this teaches us anything, it is not to fear GenAI but to understand it and use it as an advantage. And if you are still not sold, I confirmed this with ChatGPT itself. I asked, 'Will ChatGPT take our jobs?' Its answer was, 'ChatGPT will not take our jobs, but rather it will change the nature of some jobs, requiring workers to develop new skills and adapt to new ways of working.' Whether you are ready to get on board with AI or not, it is undoubtedly here to stay. ChatGPT is just one of the ways new technologies will advance our lives, jobs, and industries moving forward. Whatever the future holds, one thing is clear: The only constant is change, and as change management professionals, we are uniquely positioned to embrace it.

Further Reading

Artificial intelligence and a new era of human resources. (2023, October 9). IBM Consulting. https://www.ibm.com/blog/artificial-intelligence-and-a-new-era-of-human-resources/

Brynjolfsson, E., Li, D., & Raymond, L.R. (2023, November). *Generative AI at work* [working paper]. National Bureau of Economic Research (NBER). https://www.nber.org/papers/w31161

Eloundou, T., Manning, S., Mishkin, P., & Rock, D. (2023, August 21). *GPTs are GPTs: an early look at the labor market impact potential of large language models* [working paper]. OpenAI OpenResearch. arXiv:2303.10130v5 [econ.GN] https://arxiv.org/abs/2303.10130

Feinzig, S., & Guenole, N. (2018). *The business case for AI in HR.* IBM Consulting. https://research.gold.ac.uk/id/eprint/33662/

Vallance, C. (2023, March 28). *AI could replace the equivalent of 300 million jobs.* BBC. https://www.bbc.com/news/technology-65102150

CHAPTER 10

People Analytics Accelerates Change

by Patrick Gallagher

In the earliest days of change management, a project was either on track or it wasn't. Milestones were met or missed. Change professionals knew little about what was happening to the people undergoing a change. With advanced communication technologies during the 2000s and 2010s, we began to do better. Now, the advent of people analytics has opened further doors to understanding the effects of change on people. In this chapter, Patrick Gallagher discusses the history of measuring change, the history of change surveys, and new tools from the world of people analytics to improve change management effectiveness.

Major changes in the world of work over the last few years have given rise to new trends in talent and people strategies. Talent shortages, lowering of job switching costs with remote work, and new expectations from workers have all pushed human resources departments to invest more in data and analytics capabilities. These newer capabilities are used to measure candidate and employee sentiment, generate workforce planning predictions, and uncover the root causes of people challenges.

Bringing analytics to bear on people operations is called people analytics. It is a discipline that stretches back as far as the early 2000s, but disruptive changes in the world of work since 2020 have spurred interest. Human resources departments in companies of all sizes and in all industries have started or grown people analytics teams to gather people data, blend that data with data streams from around the enterprise, and perform analyses that uncover insights. This raises the possibility of bringing data to bear on questions that used to be answered with only personal experience or gut feelings.

Change leaders should see this evolution as a golden opportunity, as a rise in people analytics enables new ways to track and measure change, and new ways to gather feelings and feedback from stakeholders. More importantly, it has centralized resources that were historically scattered and inaccessible to change managers. Accessibility of business performance data, and the ability to blend it with employee listening data that people analytics has brought about, open new doors to expand and improve tracking progress and outcomes of change initiatives.

Historic change measurement

Change professionals cannot operate in the dark. They measure progress, attitudes, and results to plan their efforts, understand if their efforts are working, and to course correct if indicated. Measuring whether communications are being heard and how employees feel about a change helps leaders explain the change better to stakeholders. Understanding where changes are being adopted or not helps allocate scarce resources to resistant or lagging groups. And finally, the long-term outcomes of changes are monitored to help sustain benefits and calculate a return on investment (ROI) for change.

After a brief summary of the past and current landscape of change measurement, we look at improvements that people analytics have the potential to realize.

Standard approaches to measuring change progress and risk

In many cases, concrete metrics tell a clear organizational story. Switching from a legacy software system to a new one typically has benchmarks for installation and activation tracked to ensure an implementation is on schedule, for example. Utilization metrics show how many people are using a new system and how much, and even which features appear harder to transition. Data analyses not only indicate transition status, but also help recognize less-smooth components of a change, and/or which groups or individuals are not making the switch at all.

Some change efforts lend themselves to easier measurement than others. Transitioning from one SaaS (software as a service) platform to another, for example, should naturally produce data that indicates change. Other initiatives yield similarly straightforward metrics tracked with relative ease—changing signage on physical locations or transitioning from one website brand color scheme to another, for example. Some organizations, however, do not articulate what metrics are best to measure and may not be systematic in laying out plans for gathering and reporting data as initiatives progress.

A FEW COMMON METRICS FOR TRACKING CHANGE

Metric	Examples	Pros	Cons
Progress against plan	• Milestones reached on time • Old systems sunsetted on schedule • Divisions or departments reached on schedule	• Easy to measure and report • Easy for stakeholders to interpret	• Original plan might have been unrealistic or misinformed • Priorities and thus timelines can change
User adoption	• Number of users transitioned • Error-free usage of new tool • Mentions of new concepts in communications traffic	• (Generally) easy to measure and report • Easy for stakeholders to interpret	• Doesn't indicate efficiency/ effectiveness • Doesn't indicate sentiment or collateral impacts
Physical change	• Facilities closed down/opened up & operational • Physical changes (new signage) • Presence of artifacts (posters, swag, decoration)	• Concrete, obvious results • Easy for stakeholders to interpret	• Could require extensive effort to measure • Doesn't indicate sentiment or collateral impacts
Knowledge	• Awareness of the change • Understanding/ knowledge of the new • Number and content of helpdesk tickets	• Deeper understanding of the change process • Can inform ongoing communication strategy	• Requires valid survey and/or text analysis science • Doesn't indicate sentiment or collateral impacts
Sentiment	• Agreement with rationale on survey items • Most prevalent emotional responses to the change • Engagement & other attitude changes	• Deeper understanding of the change process • Can inform ongoing rollout strategy & future change	• Requires valid survey and/or text analysis science • Harder for stakeholders to interpret

© 2024, Patrick Gallagher • Future of Change Management, Gibbons & Kennedy

Figure X.1: Five dimensions of change metrics
Change metrics help change leaders know whether what they are doing is working, and how they might have to adjust.

Measuring head, hearts, knowledge, and sentiment

Change professionals know that concrete metrics do not tell the whole story of change success. How stakeholders feel about the change, how much information they're taking in, and their perceptions about its success are important to know.

For this reason, many change practitioners measure people's feelings about and reactions to a change. The subjects of this type of measurement are summarized in two categories: Knowledge and sentiment.

Knowledge

Measuring **knowledge** usually takes the form of asking workers if they are aware of a change effort, if they know details about a change, and/or if they know how to use new ideas or a new system. For example, during a culture transformation, workers might be asked if they know the new values that leadership has been promoting as part of the transformation.

> ***Editorial comment:*** Although James Healy demonstrates in another chapter that understanding values without behavioral evidence of living the values is problematic, knowledge-survey value exists when these factors are congruent.

Leaders can collect similar data on knowledge of any type of change, and analytics can reveal which components of the change are understood or which people and teams lag behind. These data can be collected in survey form, either directly asking or quizzing users about their knowledge or indirectly gathered by analyzing help requests or helpdesk tickets (or similar). The results of such analyses should help guide actions, such as what to emphasize in education or training efforts.

Sentiment

Measuring **sentiments** is typically done in similar ways—workers are asked how they feel about a change. Change leaders may directly ask workers how positively or negatively they feel about a new system or about their perceptions of how others are adapting. For initiatives that aim to change something more ephemeral like culture or employee experience, sentiments might be the main metrics to tell the story of change success. (The appropriateness of that approach depends upon the goals of the change initiative, but also on the validity and fidelity of the survey instrument.)

Survey types—what they do and don't deliver

The dominant method for gathering data on workers' reactions to change is self-report surveys. This makes sense; it is simple, direct, and familiar.

Surveys are an established method for gathering data. When they are built according to the highest scientific standards, they gather valid, insightful information about psychological processes and perceptions behind key outcomes. But surveys take several forms, and the differences between these forms are important.

Enterprise-wide employee surveys

The traditional—and probably most common—form that surveys take is the **annual enterprise-wide employee survey**. This is usually a long, comprehensive, and heavily promoted survey that everyone in the organization is encouraged or expected to complete. When designed well, this type of survey is valuable for several purposes. However, for tracking reactions to change, it is quite limited.

Annual surveys provide only a snapshot of data, frozen in time. Knowledge and sentiment regarding change, however, evolves quickly, weekly or even daily. Annual surveys also typically have limited space for the questions that change leaders are most interested in—annual surveys are often crowded with dozens of questions from many stakeholders. This makes it hard to get **specific** information on the details of **ongoing** change.

Pulse surveys

Pulse surveys—an extension of the annual survey—happen more frequently and typically have far fewer questions than annual surveys. They aim to get a "pulse" on current worker thoughts or sentiments, and typically focus on recent events or check in on engagement levels and other variables. More often, they are effective for tracking change than an annual survey, especially when they are purpose-designed to help with a specific change effort.

Pulse surveys, however, often suffer from some of the same limitations as the annual survey: When they are enterprise-wide, they do not focus on specific areas or teams where a change effort may be centered. They also can be crowded with questions from multiple stakeholders and do not allow for thorough measurement of change variables. Even though they are more frequent, they may only be semi-annually or quarterly, which does not allow for ongoing, fine-grained monitoring.

Quick targeted surveys

A third method of surveying stakeholders for their reactions to change is to deploy **quick, targeted surveys** to those affected by change, when they're affected. This method is best suited to specifically measure the effects of organizational change—it is purpose-built to measure exactly what needs to be known to optimize change and capture the information in (essentially) real-time. This type of surveying often takes the form of one or a few questions immediately after a specific experience (for example, right after a user tries a new software system) or every day or week. Multiple timepoints allow for tracking trends, to map reactions at different stages of the change effort and in different teams and departments. Where knowledge gaps or negative sentiment are detected, corrective action can be taken immediately. Many organizations lack the agility, infrastructure, or skills needed to deploy this kind of survey with change initiatives.

With all surveys, the design of the survey dictates both the kind and quality of information gathered and which analytics can be performed on the data. The more purpose-built the survey to support the change, the more effective the data are for helping.

The way most surveys are conducted, and the purposes for which they are deployed, limit their usefulness for change measurement. When change leaders want purpose-built surveys to track their work, they are often blocked for fear of "survey fatigue" or because of resource constraints. People-analytics teams are often the hub of survey activity in an organization and help determine the best timing and placement of surveys.

Barriers to better change measurement

Despite the many benefits of measuring change, it often doesn't happen. When it does, it often relies on co-opted performance metrics or a few questions on an annual employee census survey, both of which were usually designed for other purposes.

Why is this the case? It could be that decades ago, when the most popular models of change were formulated, useful data were harder to gather and analyze (this goes for all types of data and measurement efforts in organizations, not only organizational change). Because these popular models were built in circumstances without regular access to relevant data, they had little to say about measurement. Neglect of measurement remains as an artifact of those times with the change models that still dominate today.

Recently, with digitization and data competency spreading in many organizations, some have built more sophisticated, data-driven techniques to track change (e.g., Deloitte's Transformation Intelligence uses a proprietary system called Change Scout.) These techniques and systems make use of more data to measure and understand people's reactions and openness to change, and blend people and performance data.

However, these systems are often burdensome and difficult to implement. Data access remains a barrier—getting the data needed to carry out sophisticated measurement plans, and getting them in clean, analyzable form, is a challenge. It adds a burden to any change initiative to set up measurement tools and data sources—most organizations spend precious planning resources on preparing for the change and lose interest in or run out of bandwidth for measurement.

People analytics solves many of these problems for the change manager.

People analytics are a boon for change

The individual techniques and methods that makeup people analytics are not necessarily innovations in their own right. Many have existed for years, and they have been used by researchers and some practitioners to generate insights into organizational challenges. Some have even been applied to change measurement, albeit only in limited instances.

What does the rise of people analytics add to the landscape of change measurement? There are two main benefits that are game changers: Understanding and infrastructure.

Understanding

The proliferation of people analytics has brought a set of analysis and data-gathering techniques together under an umbrella set of tools. Now, more executives and business units understand the benefits of bringing data together for analyses. Increasingly, this makes "people analytics" a common shorthand that stakeholders immediately understand and endorse. In turn, the umbrella of people analytics is applied to change efforts.

One major benefit of this new shared understanding is that it also facilitates the use of behavioral science in change management.

Editorial comment: Amen.

The psychology of behavior change—at the core of any organizational change—includes habits, behavioral economics, judgment and decision-making, and social psychological principles (among other research streams). The **best people-analytics systems and teams blend behavior science variables with data from all around the organization** to enable deep insights into the drivers of change acceptance or resistance, along with fast testing of science-based interventions to accelerate change.

Infrastructure

People analytics gathers data sources together that weren't previously combined. Teams have already sourced data from around the organization and brought them into central warehouses where they are integrated and more easily analyzed. This integration is often one of the biggest barriers to effective analytics—gathering data from legacy systems, understanding them, and making them analyzable is often a major task. Change teams often do not have the data engineering skills to source and process disparate data streams.

Many people-analytics teams have done this work, so employee listening data, business intelligence from multiple departments and systems, customer data, and even data from external sources are easily accessed and analyzed. People-analytics teams often bring together employee survey data, traditionally siloed in the human resources domain, with other data from around the organization. Now, when a change initiative rolls out, its leaders may no longer have to improvise data-gathering efforts or settle for weak metrics that were designed for other purposes.

Next, we elaborate upon these developments, describe some specific tools in the people-analytics family, and illustrate the possibilities with example case studies.

Better than surveys

Natural language processing (NLP)

One of the tools that most people-analytics team deploy is **natural language processing (NLP)**, which uses algorithms to extract meaningful insights from open-text data, such as open-ended (write-in) survey responses, transcribed speech, and/or communication content.

Early forms of NLP simply counted the frequency of words in a block of text, producing the familiar word cloud graphic. Newer forms of NLP count groups of words and discard functional words that are thought not to contribute to meaning or include measures of negative or positive emotion, so the insights extracted are more contextualized and interpretable. Some sophisticated NLP tools are built on evidence-based research programs that correlate subtle language patterns with an array of psychological variables, yielding insights from text that rival or eclipse traditional surveys.

Recent advances in large-language models (LLMs) have opened up possibilities for far more advanced, automated analysis of open-text data. Tools built on LLMs can digest open text, extract themes and sentiments, and summarize with depth. Soon, widely available products will utilize such advanced capabilities, making it easier for users all over the enterprise—including change leaders—to extract insights from unstructured data.

NLP helps change leaders identify reactions among the workforce that might be hard to report on surveys, either because of biased responses, the tedium of completing surveys, or simply because workers might find it hard to articulate certain issues. To gauge interesting variables—like distress, pessimism, resolve, or delight—on a survey, valid item sets need to be identified and designed. A valid scale for each variable is three or more items long, so measuring multiple variables can quickly turn into a long, cumbersome survey for stakeholders to complete.

NLP analyses do not rely on self-report survey responses. They do, however, rely on having enough open-text data. These data could be gathered through surveys with open-ended questions or gathered through existing channels, such as email or messaging traffic.

One drawback of this method is that it invokes data privacy and ethics concerns. Many organizations feel comfortable analyzing de-identified, aggregate-language data in this way, but others might not. People analytics teams have often already navigated these issues and produced guidelines and protocols for using NLP.

LLM-based products will soon make it easier for any user to process unstructured data, meaning that change teams may not need to go through a people-analytics team or other technical owner to get insights. This, however, raises the same and additional ethical concerns—legal and/or human resource professionals should be involved in how analytics products are used to interpret data and inform decisions. The change leader may feel that deep insights are valuable and even game changing, but if they are based on flawed algorithms or biased training data, the consequences of using them are damaging.

Organizational network analysis (ONA)

Another potentially powerful change tool in the people analytics toolbox is **organizational network analysis (ONA)**. It describes the network of connections in a team or entire company, typically through measuring the frequency and/or richness of communication patterns between individuals. This allows for identification of people who may be centrally located as information "hubs," but also track the flow of information or even behaviors as they spread through an organization.

EXAMPLE ONA OUTPUT

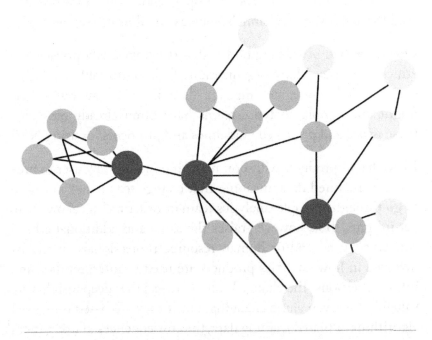

Figure X.2: Illustrative output of organizational network analysis (ONA).
An analysis of the frequency and volume of email, Slack, and other communication channels reveal patterns like those depicted. Darker circles represent people who are particularly central or could potentially bridge silos, making them strong candidates to help lead change adoption.

ONA is conducted either using surveys (essentially by asking people to report who they follow for information or communicate with the most) or through analysis of other business data. This can include email or messaging traffic, which allows for identifying different individuals who communicate with many or few others measuring the frequency, length, and/or content of those connections. Phone calls, posting and liking activity on internal forums, or any other data that captures reach and/or content of communications can also be used in this way.

One way of using ONA, which has already found its way into change efforts, is to identify those workers who might act as change champions by influencing their extensive network of connections. If a centrally located source of guidance or information adopts a new system or new process early on, they might quickly and positively influence others to do so.

Use case example: ONA in healthcare change

To reach their goal of becoming the premier nursing education center, Duke University School of Nursing (DUSON) needed systematic organizational transformation. As part of the design of this change effort, the school conducted an ONA to identify the existing networks of communication and work contacts within the organization.

This analysis informed all aspects of the change initiative design. They structured its project steering committee and its rollout of structural changes to maximize the influence of change champions and to take advantage of how work was already being performed. The ONA "...gave me a natural source of leaders in addition to the formal leadership structure," said the DUSON's leader, which then helped effectively roll out the changes.

Another use of ONA is to track behavior spread—for example, to identify patterns of uptake for a new system. Such an analysis identifies teams or departments that are early adopters, and then tracks whether and where those behavior changes spread over time. Such analyses uncover patterns of spread that might not depend on explicit communication channels, but instead on overlapping processes or system dependencies. They might also uncover areas of friction and predict where change adoption might hit a barrier because of unexpected or unwanted process changes in affected teams.

NLP and ONA are just two tools in the people-analytics toolbox that facilitate smarter and faster change. As already mentioned, surveys are another: Survey deployment and blending employee listening data with performance data help change leaders identify sub-populations of early adopters, track emotional reactions, and understand changes' impact on bottom-line metrics.

Predictive modeling

Another people-analytics tool to aid change efforts is predictive modeling. People-analytic teams often include data scientists who build models of important outcomes based on historical data. For example, people-analytics teams often predict turnover rates in certain departments so human resources' teams can plan talent sourcing and hiring to maintain optimal staffing levels. Models may also attempt to predict benefits utilization, the total cost of expanding or contracting a workforce, and other valuable outcomes.

The success of predictive modeling relies on the quality of its input data—that is, a very large set of historical data is more likely than a small set to yield reliable predictions for the future. Even with adequate data, however, predictive models are rendered useless when an unanticipated circumstance changes conditions. For example, no predictive workforce models in 2019 included a pandemic and economic shutdown (and those that included this scenario were probably not acted upon).

Nevertheless, a talented people-analytics team could build predictive models to help a change effort plan resources and strategies. For example, modeling different uptake rates for a new process flow, helping a change team prepare for several scenarios in terms of cost, communication resources, and/or IT support needs. Blending behavioral science knowledge and a change leader's expertise into these models have potential to yield helpful predictions that benefit change efforts and the organization.

Beyond the scenarios described above, there are surely additional unexplored possibilities for using the people-analytics toolkit to strengthen change efforts. No matter what form those possibilities take, change teams should lay the groundwork for people analytics collaboration.

The following are some tips for getting started.

People analytic tips for change success

Following are six ways (summarized in Figure X.3) that organizations can bring the benefits of analytics to change efforts. Human-resource teams (e.g., survey/employee experience teams or learning-and-development teams) and analytics-and-data teams (e.g., data engineers, data scientists, data warehouse teams) can collaborate to help with gathering and analyzing data.

SIX STEPS TO USING PEOPLE ANALYTICS FOR CHANGE SUCCESS

1	Get to know your PA team!	Find out who's doing PA in your organization, and start collaborating on generate value.
2	Listen for change	Work with PA to design listening channels that will yield directly relevant change information.
3	Define metrics – and track them	Find the metrics that matter to your goals. Ensure the PA team gathers and track them.
4	Measure and analyze quickly	Set up processes that support quick turnaround – you'll need the agility.
5	Find people with evidence	Your impressions of who leads change may not be accurate. Follow the data.
6	Translate	Perhaps the hardest, and most important, task is making something happen.

© 2024. Patrick Gallagher • Future of Change Management, Gibbons & Kennedy

Figure X.3: Six tips for using people analytics to aid change efforts.
Enlisting people analytics to help with change efforts may not be easy, but the change leader who follows these steps may gain efficiency and effectiveness.

Get to know your people analytics team!

This seems an obvious first step. Change leaders should reach out to people doing people analytics at their organization. Get to know their capabilities, the kinds of work they do, and the tools they've built. Engage them in conversations about how their efforts could combine with change initiatives to accelerate the value of both.

Many people-analytics teams have been built recently and now face the challenge of demonstrating the value they return to an organization. Facilitating change through measurement provides a profitable use case.

Design listening channels

People analytics teams will typically have extensive experience with designing and strategically deploying surveys, as well as other data streams used to hear employees' feedback and concerns. They may be able to supply open-text data for analysis of change-related variables. They can also help target surveys to assess the variables most important to change teams and deploy them only to those teams where they are needed most. They can likely also weave them into established rhythms of data collection so that new surveys are not seen as new burdens.

People analytics can also potentially blend those employee listening data with performance data and run NLP and ONA, among other techniques, in real time. Instead of taking snapshots of data, merging them manually, and analyzing for point-in-time results, the best people-analytic systems enable insights to be tracked as they unfold. All this has the potential to make survey data far more powerful for change initiatives.

Define metrics, then revisit them

How will you know if your change efforts are successful? What reactions, learning, or behavioral outcomes need to increase or decrease? These are the questions that lead to identifying the best metrics to track. If those metrics are well-defined and understood, then people-analytic teams can source them, make sense of them, and generate useful insights from them. Counting how many people are exposed to a change intervention is an unsophisticated metric, measuring changes in attitudes or knowledge is better, and tracking key behavior changes in the central actors is better still. People-analytic systems are built to gather such data and even compare treatment and control groups, again in real time, which help isolate the effects of change interventions among the noise of other business events.

Importantly, you may not be best served by keeping the same metrics indefinitely. If you discover that certain leading indicators aren't in fact predicting the outcomes you thought they would, then consider abandoning them. Likewise, if you discover a predominant theme in workers' reactions to an intervention that you didn't anticipate, don't hesitate to begin tracking it. The goal is to maximize benefits of the change, not keep tracking certain metrics for their own sake. Validate your metrics by ensuring that they are telling you what you need to know. If you see that validation begin to wane, revisit those original questions about how to know if you're being successful, and revise. A good relationship with people analytics partners enable this agility.

Measure frequently and analyze quickly

If your measurement and analysis cycle is quick, you can identify issues and course correct before incurring much loss or damage. You can also identify things that are working well, and generalize them, thereby accelerating successful change. Having real-time data collection, cleaning, and analysis set up enable quick reads on adoption, knowledge, and outcomes of changes.

Getting this right takes more than data engineering or analytics skills, it takes thoughtful planning and agility on the change team. Keep in mind that tracking and analyzing the data you want might be time-consuming and difficult, even for a well-established people-analytics team. Simple metrics and analyses that don't depend on scarce resources (but still track the most important metrics) might end up enabling more analytics than a complex set of metrics and dashboarding.

Identify change agents with evidence, not only impressions

Your or others' impressions of who the most influential people are might be right, or they might not. Anecdotally, I've seen many analytics results point to conclusions that leadership or others in an organization already recognized in some form. However, the data reveal other insights as well. There may be a quiet influencer, who does not cut a high profile at meetings or in the office, but regularly distributes emails with valuable information to those who need it. These are individuals who might be effective change champions, but don't fit the prototypical image. ONA, or other data gathered to track change efforts, might reveal central people or teams that are playing roles you didn't expect. Watch for those dynamics!

Cover the last mile—translation

Analytics don't help if data are gathered but not translated into action. Getting data collection and analysis mechanisms in place is a big task, but building the skill, insight, and creativity to translate results is as tricky.

Although the best people-analytics teams provide end-to-end service (from data identification and sourcing to finished reports), many teams will likely only produce descriptive statistics that do not directly prescribe action. Your change team will likely need to interpret those outputs and cover the last mile of translation—figuring what actions the numbers support. This may require some deep logic—what data points are actually moving the needle on interesting outcomes? What outcomes are really the most important? What lies underneath the descriptive statistics in the dashboard—for example, what are the top drivers of resistance in lagging areas? Creative analysts, behavioral scientists, and skilled data engineers (or an agile external provider of data engineering) are needed to identify and then answer these questions. Those who understand the output of people analytics and also have on-the-ground experience of getting things done are in short supply.

Conclusion

Change teams have been using analytics, in primitive forms, for decades. But there is more value to be found in analytics than most organizations currently get. Sophisticated use of surveys and advanced techniques, like ONA, give leaders and practitioners better information about changes and their impact on stakeholders. Blending survey and other business intelligence data also have potential to make change efforts more efficient and effective. The rise of people analytics, and the resulting availability and usability of new data types, may help change leaders realize the full promise of analytics in change management.

Further Reading

Cullen-Lester, K., & Willburn, P. (2016). *Analytics for change: how networks and data science will revolutionize organizational change*. Center for Creative Leadership. https://cclinnovation.org/wp-content/uploads/2020/02/analyticsforchange.e.pdf

Duke University School of Nursing: leveraging network analysis for change center for creative leadership. (n.d.). Center for Creative Leadership. https://www.ccl.org/client-successes/case-studies/duke-university-school-of-nursing-leveraging-networks-for-change/

Evaluating engagement and mental wellness of remote workers through email analysis. (2021, September 1). Receptiviti. https://www.receptiviti.com/post/evaluating-engagement-and-mental-wellnessof-remote-workers-through-email-analysis

Ferrar, J. & Green, D. (2021). *Excellence in people analytics: how to use workforce data to create business value* (1ˢᵗ ed.). KoganPage.

Gibbons, P. (2021). *Data-driven change management using Transformation Intelligence*™. Deloitte. https://www2.deloitte.com/content/dam/Deloitte/us/Documents/human-capital/us-data-driven-change-management-using-transformation-intelligence.pdf

Khan, N., & Millner, D. (2020). *Introduction to people analytics: a practical guide to data-driven HR* (1ˢᵗ ed.). KoganPage.

CHAPTER 11

Conclusion

by Paul Gibbons and Tricia Kennedy

"Well begun is half done."
ARISTOTLE

With such a dispersed array of topics, we resist the temptation to tie a pretty bow around them all—knitting together the micro (of say neuroscience) and the meso (of behaviors) to macro-level tools (such as design thinking.)

But these are the different levels at which the skilled change practitioners must operate. From understanding (a little) about brains, to being able to link that understanding with research on behaviors, to having tools that allow them to operate at "the system" level. That is what makes the job fun and interesting – well, most days.

We learn more about the "micro" level of organizational change—the section we call "minds," every year. This is an important piece of a "humanizing change" project birthed by Paul and Tricia some five years ago. For change management to become more human, it must hew more closely to the sciences of mind, such as psychology, sociology, and cognitive-affective neuroscience.

CONCLUSION

Paul remembers one shocking finding, from his masters' degree program in organizational behavior, 30 years ago: "10% of managerial training transfers back to the job.[1]" Shocking because training and education are a go-to tool for changing behavior in organizations—say cybersecurity, discrimination, or technology adoption. (Moreover, businesses spend $100 billion give or take on this stuff. Ouch.) If that finding is even approximately right, we are simply terrible at changing human behavior through training. (And you might ask yourself how much of that course on time management you still use.) Similar research indicates that persuasion and influencing produce results just as paltry. (Ask yourself how often you've persuaded someone that a course of action is a good idea, they've agreed, and then done nothing differently.) For that reason, this book has three chapters on behaviors as we believe behavioral science can begin to close some of that gap—between intentions and actions, between head and hands.

Robert Meza's chapter on behavioral science tools (chapter 7) could have gone either under behaviors or tools. We chose the latter having heard many times "this behavioral science stuff is fascinating and sure seems to work way better than traditional behavioral change approaches, but it is hella' complex, how do I use it?" Chapter 7 is a start and step in the right direction. We are similarly excited about the design thinking, ChatGPT, and people analytics chapters as practical change tools. While some change experts are familiar with some of these topics, fully integrating them into our ways of working (models, tools, frameworks, interventions, etc.) has some way to go.

Volume 2 in the works

As promised in the introduction, a second volume is in the works. Conversations with potential contributing authors are underway.

Please contact the editors with any topics you would especially like to see included, or if you would like to write a chapter.

Two topics we wanted in volume 1 were **psychological safety** and **diversity, equity, and inclusion (DEI)**. We aim to rectify that in the next volume.

Also on our radar is the burgeoning world of **change management software**—in our estimation, there are dozens of companies offering such solutions. What ground do they cover and what problems do they solve? What are the risks and drawbacks? What human change management activity can they legitimately sub for? (Note, one such company loudly promotes itself as "the future of change management." OK, sure buddy – whatever.)

Lastly, **GenAI and organizational change** was a brand-new topic with a brand-new technology when Natasha began her experiments with ChatGPT. Since then, much has changed. The transformer, LLM, and GenAI world is white hot with new developments, and many consultants and academics have broached this topic. We aim to have at least one contribution in this area in the next volume.

CONCLUSION

VOLUME 2 CANDIDATE TOPICS

✓ Diversity and inclusion
✓ GenAI
✓ Psychological safety
✓ **Change management software**
✓ **Evidence-based change**
✓ Change leadership 2030
✓ Remote work and change

• Future of Change Management, Gibbons & Kennedy

Figure XI.1: The Future of Change Management (volume 2).
Candidate topics under consideration for the next, or second, volume in The Future of Change Management series.

"The Moving Finger writes; and, having writ,
Moves on: nor all thy Piety nor Wit
Shall lure it back to cancel half a Line,
Nor all thy Tears wash out a Word of it."
OMAR KHAYYÁM

APPENDIX I
Contact the contributing authors

Contributing authors are listed in alphabetical order.

Beirem Ben Barrah beirem@neurofied.com

Newton Cheng newtoncheng@gmail.com

Ignacio Etchebarne ignacio@hi-web.com.ar

Patrick Gallagher patrick4440@gmail.com

Paul Gibbons, paul@paulgibbons.net

James Healy jahealy@deloitte.com.au

Philip Jordanov philip@neurofied.com

Tricia Kennedy TK@triciak.com

Robert Meza hi@aimforbehavior.com

Hilary Scarlett hilary@scarlettandgrey.com

APPENDIX I

Yves Van Durme yvandurme@deloitte.com

Natasha Young natasha.j.young@outlook.com

Scott Young sy28mackay@gmail.com

References and Further Reading

(alphabetical order)

Armstrong, T. (2010). *The power of neurodiversity: unleashing the advantages of your differently wired brain.* Hachette Books. Kindle edition.

Arnsten, A.F.T. (2009). Stress signaling pathways that impair prefrontal cortex structure and function. *Nature Reviews Neuroscience, 10*(6), 410-422.

Artificial intelligence and a new era of human resources. (2023, October 9). IBM Consulting. https://www.ibm.com/blog/artificial-intelligence-and-a-new-era-of-human-resources/

Ben Barrah, B., & Jordanov, P. (2024). *The dynamics of business behavior: an evidence-based approach to managing organizational change* (1st ed.). Wiley.

Brown, T. (2008, June). *Design thinking.* Harvard Business Review.

Brynjolfsson, E., Li, D., & Raymond, L.R. (2023, November). *Generative AI at work* [working paper]. National Bureau of Economic Research (NBER). https://www.nber.org/papers/w31161

Cullen-Lester, K., & Willburn, P. (2016). *Analytics for change: how networks and data science will revolutionize organizational change.* Center for Creative Leadership. https://cclinnovation.org/wp-content/uploads/2020/02/analyticsforchange.e.pdf

Culture eats strategy for breakfast. (2013, May 17). Quote Investigator. https://quoteinvestigator.com/2017/05/23/culture-eats/

REFERENCES AND FURTHER READING

Dawson, P., & Guare, R. (2016). *Smart but scattered. Guide to success. How to use your brain's executive skills to keep up, stay calm, and get organized at work and at home.* Guilford Publications. Edición de Kindle.

DeBellis, P. (2020, December 2). *Adaptable by design: a future-focused, fit-for-purpose HR operating model.* Deloitte. https://www2.deloitte.com/us/en/blog/human-capital-blog/2020/hr-operating-model.html

Dunbar, R., Camilleri, T., & Rockey, S. (2023). The social brain: the psychology of successful groups. Penguin Books UK.

Duke University School of Nursing: leveraging network analysis for change center for creative leadership. (n.d.). Center for Creative Leadership. https://www.ccl.org/client-successes/case-studies/duke-university-school-of-nursing-leveraging-networks-for-change/

Eloundou, T., Manning, S., Mishkin, P., & Rock, D. (2023, August 21). *GPTs are GPTs: an early look at the labor market impact potential of large language models* [working paper]. OpenAI OpenResearch. arXiv:2303.10130v5 [econ. GN] https://arxiv.org/abs/2303.10130

Etchebarne, I. (2022). *Cuando la neurodiversidad excluye. 6 supuestos erróneos del movimiento de la neurodiversidad.* LinkedIn [Post]. https://www.linkedin.com/pulse/cuando-la-neurodiversidad-excluye-ignacio-etchebarne/

Evaluating engagement and mental wellness of remote workers through email analysis. (2021, September 1). Receptiviti. https://www.receptiviti.com/post/evaluating-engagement-and-mental-wellnessof-remote-workers-through-email-analysis

Feinzig, S., & Guenole, N. (2018). *The business case for AI in HR.* IBM Consulting. https://research.gold.ac.uk/id/eprint/33662/

Ferrar, J. & Green, D. (2021). *Excellence in people analytics: how to use workforce data to create business value* (1st ed.). KoganPage.

Forsythe, J., Duda, J., Cantrell, S., Scoble-Williams, N., & Marcotte, M. (2024). One size does not fit all—Deloitte human capital trends 2024. Deloitte Consulting. https://www2.Deloitte.com/xe/en/insights/focus/human-capital-trends.html#one-size-does-not-fit-all

Gibbons, P. (2021). *Data-driven change management using Transformation Intelligence*™. Deloitte. https://www2.deloitte.com/content/dam/Deloitte/us/Documents/human-capital/us-data-driven-change-management-using-transformation-intelligence.pdf

Gibbons, P., & Cheng, N. (2024, January). *Mental health and work (guest Newton Cheng)*. Think Bigger, Think Better podcast.

Gibbons, P., & Hollon, S. (2019, August). *Depression, how can you spot it? What can you do? (guest Steve Hollon)*. Think Bigger, Think Better podcast.

Gibbons, P., & Kennedy, T. (2023). *Change myths: the professional's guide to separating sense from nonsense*. Phronesis Media: Kindle edition.

Gibbons, P. (2019). *The science of organizational change: how leaders set strategy, change behavior, and create an agile culture* (2nd ed.). Phronesis Media.

Gollwitzer, P.M. (1999). Implementation intentions: strong effects of simple plans. *American Psychologist, 7*(54), 493.

Grant, A. (2014). *Give and take*. Orion Publishing.

Guiso, L., Sapienza, P., & Zingales, L. (2013, October). The value of corporate culture. National Bureau of Economic Research (NBER). https://www.nber.org/system/files/working_papers/w19557/w19557.pdf

In first person: Satya Nadella. (n.d.). People + Strategy Journal, Society of Human Relations Management (SHRM). https://www.shrm.org/executive/resources/people-strategy-journal/fall2020/pages/in-first-person.aspx

Khan, N., & Millner, D. (2020). *Introduction to people analytics: a practical guide to data-driven HR* (1st ed.). KoganPage.

Klein, G. (2007). Performing a project postmortem. *Harvard Business Review, 9*(85), 18-19.

Lilienfeld, S., & Arkowitz, H. (2015). *Facts and fictions in mental health*. Wiley.

Maslach C., Leiter M.P. (2016). Understanding the burnout experience: recent research and its implications for psychiatry. *World Psychiatry, 15*(2),103-11.

Mazor, A., Johnsen, G., Stephane, J., Hill, A., Calamai, J.B., & Moen, B. (2019). *Exponential HR: break away from traditional operating models to achieve work outcomes*. Deloitte.

Meza, R. (n.d.). *Behavior design tools*. https://courses.aimforbehavior.com/free-behavior-and-innovation-frameworks

Michie, S., Atkins, L., & West, R. (2014) *The behaviour change wheel—a guide to designing interventions*. Silverback.

Michie, S., et al. (2013). The behavior change technique taxonomy (v1) of 93 hierarchically clustered techniques: Building an international consensus for the reporting of behavior change interventions. *Annals of Behavioral Medicine, 46*(1), 81–95.

Michie, S., van Stralen, M.M. & West, R. (2011). The behaviour change wheel: A new method for characterising and designing behaviour change interventions. *Implementation Science, 6*, 42.

Muñoz R.F., Beardslee W.R., & Leykin, Y. (2012). Major depression can be prevented. *American Psychologist, 67*(4), 285-95.

Schnall, S., Harber, K.D., Stefannucci, J.K., & Proffitt, D.R. (2008). Social support and the perception of geographical slant. *Journal of Experimental Social Psychology, 44*(5), 1246-1255.

Service, O., Hallsworth, M., Halpern, D., Algate, F., Gallagher, R., Nguyen, S., Ruda, S., & Sanders, M. (2014). *Four simple ways to apply EAST framework to behavioral insights.* The Behavioral Insights Team. https://www.bi.team/publications/east-four-simple-ways-to-apply-behavioural-insights/

Shanafelt, T.D., West, C.P., & Sloan, J.A. (2009). Career fit and burnout among academic faculty. *Archives of Internal Medicine, 169*(10), 990-995.

Sibony, O. (2020). *You're about to make a terrible mistake! How biases distort decision making and what you can do to fight them.* Swift Press.

Solow, M., & Wakefield, N. (2016, February 29). *Design thinking: crafting the employee experience.* Deloitte. https://www2.deloitte.com/us/en/insights/focus/human-capital-trends/2016/employee-experience-management-design-thinking.html

Soman, D., & Yeung, C. (eds.) (2020). *The behaviorally informed organization.* University of Toronto Press.

Soman, D. (2020). Sludge: a very short introduction. Behavioral Economics in Action at Rotman, Nudgestock annual conference. https://www.rotman.utoronto.ca/-/media/Files/Programs-and-Areas/BEAR/White-Papers/BEARxBIOrg-Sludge-Introduction.pdf?la=en&hash=DCB98795CB485977A04DDB27EFD800C3DA40220E

Southwick, S. M., & Charney, D. S. (2012). *Resilience: the science of mastering life's greatest challenges.* Cambridge University Press.

Stroebe, W., & Diehl, M. (1994). Why groups are less effective than their members: on productivity losses in idea generating groups. *European Review of Social Psychology, 1*(5), 271-303.

Sull, D., Sull, C., & Chamberlain, A. (2019, June 24). Measuring cultures in leading companies. MIT Sloan Management Review. https://sloanreview.mit.edu/projects/measuring-culture-in-leading-companies/

The 20 most influential business books. (2002, September 30). Forbes. https://www.forbes.com/2002/09/30/0930booksintro.html?sh=55a0536c29e1

The Inglehart-Welzel World Cultural Map. World Values Survey 7 (2023). Source: http://www.worldvaluessurvey.org/

Torralva, T., Gleichgerrcht, E., Lischinsky, A., Roca, M., & Manes, F. (2018). "Ecological" and highly demanding executive tasks detect real-life deficits in high-functioning adult ADHD patients. *Journal of Attention Disorders, 17*(1), 11-19. https://doi.org/10.1177/1087054710389988

Vallance, C. (2023, March 28). *AI could replace equivalent of 300 million jobs.* BBC. https://www.bbc.com/news/technology-65102150

van Bavel, J., & Packer, D. (2021). The power of us: harnessing our shared Identities for personal and collective success. Headline.

van den Aker, M. (2023, November 19). Interview with Scott Young. Money on the Mind. https://www.moneyonthemind.org/post/interview-with-scott-young

Waterman Jr., R.H., Peters, T.J., & Phillips, JR. (1980). Structure is not organization. Business Horizons, June. https://managementmodellensite.nl/webcontent/uploads/Structure-is-not-organization.pdf

Weisbord, M., & Janoff, S. (2010). *Future search: getting the whole system in the room for vision, commitment, and action.* Berrett-Koehler Publishers.

Young, S. (2023 April 25). How (& why) to start infusing your company with behavioral science. Ethical Systems. https://www.ethicalsystems.org/how-why-to-start-infusing-your-company-with-behavioral-science/

Young, S. (2020, June 22). Finding opportunities to apply behavioral science for good in the private sector. Behavioral Scientist. https://behavioralscientist.org/finding-opportunities-to-apply-behavioral-science-for-good-in-the-private-sector/

Yerkes, R.M., & Dodson, J.D. (1908). The relation of strength of stimulus to rapidity of habit formation. *Journal of Comparative Neurology and Psychology, 18*(5) 459-82.

Made in the USA
Columbia, SC
26 December 2024